Construction Law & Risk Management

Case Notes

Volume II

by

J. Kent Holland, Esq.

Kent@KentHolland.com

Ardent Publications
8596 Coral Gables La.
Vienna, VA 22182
703-623-1932

Copyright © 2006, by J. Kent Holland, Jr.

All rights reserved. No part of this publication may be reproduced, stored in a retrieval system, or transmitted in any form or by any means—electronic, mechanical, photocopying, recording, or otherwise—without the prior written permission of the copyright owner.

Published by Ardent Publications – Vienna, VA
www.ArdentPublications.com

Author e-mail: Kent@KentHolland.com
Author Phone: 703-623-1932

ISBN 0972315837

Library of Congress Control Number: 2005910587

Disclaimer

No liability is assumed with respect to the use of the information contained herein, nor shall the publisher, distributor, author, or any employer or entity with which the author is affiliated be liable for damages or loss resulting from its use, including but not limited to actual, consequential, or incidental damages, whether foreseeable or unforeseeable. Although every precaution has been taken in the preparation of this book, the author assumes no responsibility for errors or omissions.

The opinions expressed herein are solely those of the author, and do not represent the views of any individual, partnership, corporation, institution of learning, or any other entity by which the author is or has been employed or affiliated.

This book is written, published and distributed in the author's individual capacity and is independent of any insurance company, law firm or other enterprise with which the author may be affiliated. It is distributed with the understanding that the author, contributors, publisher and distributors are not hereby engaged in rendering legal services or the practice of law. Further, the content and comments in this text are provided for educational purposes only and are for general distribution. They cannot apply to any single set of specific circumstances, and should not be applied without the review and approval of your attorney.

While this book contains explanations of various legal concepts, it is not legal advice. It is important to keep in mind that statutory and common law vary form state to state. How one court may interpret the language of contract, insurance policy, or document may differ significantly form how it will be interpreted and applied by another court in another jurisdiction. Legal advice concerning any matter presented or discussed in this book should be sought from counsel knowledgeable in the law of the jurisdiction in which services will be performed and under which the applicable contract or insurance policy will be enforced.

Preface

This book compiles and organizes case notes, articles, and papers written by well known and respected attorneys and professional consultants for original publication in a number of newsletters. Articles that do not include a specific byline were written by Kent Holland and first published in *ConstructionRisk.com Report* (www.ConstructionRisk.com), or in the International Risk Management Institute's *Expert Commentary* (www.irmi.com). The cases and articles included in this book demonstrate risk management principles to be considered and applied on construction projects. The intent is to give a sampling of issues and cases, providing risk management ideas and information to serve as a useful resource for contractors, design professionals, project owners, attorneys, educators, risk managers, and insurance professionals. Volume I covered case notes published in Volumes 1 through 4 of *ConstructionRisk.com Report*. This second volume is based primarily on case notes published in Volumes 5, 6, and 7.

Each case note title is followed by a parenthetical with the volume and issue number of the *ConstructionRisk.com Report* in which the case note first appeared. Case notes that were originally published in the *IRMI Expert Commentary* show the date and year of their original publication by IRMI.

About the Author

J. Kent Holland is an attorney with the Virginia office of a nationally recognized law firm, with a legal practice emphasizing construction, government contracts, federal grants, and insurance law. He also provides risk management services for the environmental and design professional unit of a major insurance carrier.

From 1982 through 1986, he was an attorney in the Office of General Counsel of the U.S. Environmental Protection Agency, with responsibility for assisting the Agency in deciding wastewater treatment construction grants disputes, contractor claims, bid protests, and debarment matters.

Mr. Holland is a frequent speaker on the subjects of environmental law, insurance law, and construction law, with a special emphasis on risk management for design professionals and contractors. He has written several books, including *Risk Management & Contract Guide for Design Professionals* (2006); *Construction Law & Risk Management - Case Notes and Articles*, (2003); *Architectural/Engineering Contracts Risk Management Guide* (1997); and *EPA Construction Grants Disputes: Surviving the Audit* (1990). He has also has written chapters in several manuals and books for publishers including, Wiley Law, Aspen Law, and International Risk Management Institute (IRMI).

Mr. Holland publishes a web-based construction risk management library and legal newsletter at http://www.ConstructionRisk.com. He is a 1979 graduate of the Villanova University School of Law.

Acknowledgements

Most of the material in this book was originally published in monthly issues of the *ConstructionRisk.com Report* during the years 2003 through 2005. Articles that do no include a by-line were written by Kent Holland. All articles were contributed by others include their by-line and contact information. Acknowledgement is gratefully given to the attorneys and professional consultants whose articles and case notes have been included in this book. Attribution to them and their firm, is provided with their individual articles.

A number of articles included within this book were originally written by Kent Holland for inclusion in the Expert Commentary section of the International Risk Management Institute (IRMI) Web site and newsletter. We commend the IRMI Web site at http://www.IRMI.com as an excellent and valuable resource for additional articles and construction risk management information.

Contributing Authors

Beth Andrus, Esq. – Skellenger Bender, P.S
J. Gerard Boyle – Revay and Associated Limited
Michael J. Carrato, Esq.
Andrew B. Cohn, Esq. – Kaplin Stewart Meloff Reiter & Stein, PC
Stephen J. Densmore, Esq. – Heyman & Densmore
Daniel J. Donohue, Esq. – Wickwire Gavin, P.C.
Allan H. Goodman – GSBCA Judge
M.K. Holohan, Esq. – formerly with Wickwire Gavin, P.C.
Gerald Katz, Esq. – Katz & Stone, LLP
Susan Linden McGreevy, Esq. – Husch & Eppenberger
Michael Loulakis, Esq. – Wickwire Gavin, P.C.
Daven G. Lowhurst, Esq. – Thelen Reid & Priest LLP
Robert J. MacPherson, Esq.
Bruce Marvin, Esq. – Skellenger Bender, P.S
Lawrence Moss, Esq. – D'Ancona & Pflaum LLC
Julie M. McGoldrick, Esq. – Wickwire Gavin, P.C.
Hal Perloff, Esq. – Wickwire Gavin, P.C.
Gordon Rees, L.L.P.
Gregory R. Shaughnessy, Esq. – Thelen Reid & Priest LLP
Katz & Stone, LLP
Philip R. White, Esq. – Sills, Cummis, Radin, Tischman, Epstein & Cross
Richard Zarandona – Arch Insurance Group

Chapters

1.0	Accord & Satisfaction
2.0	Americans with Disabilities Act
3.0	Changes: Managing Change Orders
4.0	Contractor Claims
5.0	Contract Language Issues & Concerns
6.0	Damages
7.0	Design-Build
8.0	Dispute Resolution
9.0	Documentation
10.0	Drug Testing
11.0	Economic loss Doctrine
12.0	Environmental Liability
13.0	Ethics
14.0	Expert Witnesses
15.0	Federal Contracts
16.0	Fiduciary Duty
17.0	Indemnification
18.0	License Requirements
19.0	Insurance
20.0	Insurance Coverage for Environmental Losses & Mold
21.0	Limitation of Liability
22.0	Mold
23.0	Site Safety
24.0	Standard of Care
25.0	Surety
26.0	Time Limitations on Suits
27.0	Warranty

Table of Contents

1.0 Accord & Satisfaction
1.1 Accord and Satisfaction Barred Contractor Claim for Additional Compensation
1.2 Accord and Satisfaction Language Barred Contract Claim
1.3 Waiver and Release in Change Order Bars Further Recovery
1.4 Do You Really Want to Cash that Check?

2.0 Americans with Disabilities Act
2.1 Design Professionals Not Subject to Liability under Title III of the ADA and Washington's Law against Discrimination

3.0 Changes: Managing Change Orders
3.1 Managing Contract Changes

4.0 Contractor Claims
4.1 Contractor Suit Dismissed for Failure to Follow Claim Procedures of Contract
4.2 Failure to Request Change Order Bars Contractor Recovery for Excess Units
4.3 Multi-Million Dollar Claim Invalidated by Court Due to Contractor's Failure to Give Timely Notice
4.4 Absent Contractual Allocation of Delay Risk, Subcontractor Cannot Recover Damages from General Contractor who was not Responsible for their Occurrence
4.5 Equitable Adjustment Allowed for Deductive Change Despite Contractor's Unbalanced Bid
4.6 Proving Constructive Acceleration Claims Can be Difficult for Contractors
4.7 When to Stop Work for Non-payment

4.8 Differing Site Conditions, Defective Specifications: One Coin, Two Sides

5.0 Contract Language Issues & Concerns
5.1 Boilerplate Can Burn!

6.0 Damages
6.1 Contractor May be Sued for Lost Profits arising out of Breach of Contract

7.0 Design-Build
7.1 Book Review: *Design-Build Lessons Learned*
7.2 Ambiguity in Design-Build Contract Specs Interpreted in Favor of Contractor
7.3 Design-Builder Not Entitled to Equitable Adjustment to Meet Owner's Detailed Design Specifications
7.4 Liquidated Damages Clause and Waiver of Consequential Damages Clause Effectively Cap Damages Available Against Design-Builder
7.5 Public Agency Exempted Project from Competitive Bidding

8.0 Dispute Resolution
8.1 Contractual Jury Waivers Held Invalid by California Supreme Court
8.2 Subcontractor forfeits right to arbitration by filing demand untimely
8.3 Using Negotiation, Mediation and Arbitration to Resolve Construction Disputes.
8.4 Why Some Mediations Fail
8.5 Arbitration Consolidation was Inappropriate
8.6 Architect's Decision Final where Contractor Failed to Satisfy Arbitration Filing Requirements

9.0 Documentation
9.1 Don't Touch That "Forward" Button! Attorney-Client Privilege in an E-Mail Age
9.2 Copyright Infringement of Design Documents

10.0 Drug Testing
10.1 Rapid Result Drug Testing

11.0 Economic loss Doctrine
11.1 Contractors May Now Bring Direct Action for Economic Losses Against Design Professionals in Pennsylvania

12.0 Environmental Liability
12.1 Superfund Decision May Benefit Design Professionals on Environmental Remediation Projects

13.0 Ethics
13.1 Testing Your Ethical Barometer

14.0 Expert Witnesses
14.1 Contractor Complaint against Engineer Dismissed for Failure to File Expert Identification Affidavit
14.2 Personal Injury Case against Engineer Dismissed for Lack of Expert Testimony

15.0 Federal Contracts
15.1 Hurricane Katrina's Impact on Existing U.S. Government Contracts

16.0 Fiduciary Duty
16.1 Project Manager Required by Fiduciary Duty to Owner to agree to Settlement with a Supplier Contrary to its Own Interest

17.0 Indemnification
17.1 Indemnity Clause Requires Subcontractor to Indemnify Prime for Injuries Arising out of the Prime's own Negligence
17.2 Highway Contractor Protected by State Immunity Statute

17.3 Indemnification Clause Unenforceable if Negligent Parties Are Indemnified

18.0 License Requirements
18.1 Contractor Forfeited Right to Payment by Performing Work without a License

19.0 Insurance
19.1 Waiver of Subrogation Enforced, Denying Insurance Company Recovery against Contractor it Alleged was Grossly Negligent
19.2 Faulty Workmanship Coverage Under CGL Policy
19.3 Pollution Exclusion in D&O Policy Applied to Exclude Coverage for Alleged Business Torts
19.4 Insurance Carrier not Required to Treat CM as Additional Insured Under Contractor's Policy
19.5 Broad Additional Insured Endorsement Entitles Contractor to Recover Damages under its Subcontractor's Primary and Umbrella Policies
19.6 Punitive Damage Award Against Insurance Company Reversed by Supreme Court as Excessive
19.7 Insurance Company that Incorrectly Denied Pollution Coverage Did not Act in Bad Faith in Failing to Defend and Indemnify its Insured
19.8 Insurance Coverage—Waivers of Subrogation

20.0 Insurance Coverage for Environmental Losses & Mold
20.1 Silica Claim Barred by Total Pollution Exclusion in CGL Policy
20.2 Broad Pollution Exclusion Is Ambiguous: Lead Covered by Policy
20.3 Whether Mold Cleanup Costs Are Covered Depends on Causation
20.4 Mold Loss Excluded under Homeowner's Policy – Summary Judgment for Carrier

20.5 Absolute Pollution Exclusion in Contractors Policy Does Not Bar Coverage for Injuries from Toxic Fumes

21.0 Limitation of Liability
21.1 Limitation of Liability Clause Protecting Owner is Not Voided by Owner's Breach of Contract or Alleged Bad Faith
21.2 Liquidated Damages Clause and Waiver of Consequential Damages Clause Effectively Cap Damages Available against Design-Builder

22.0 Mold
22.1 Standards Needed for Mold Exposure, Testing and Remediation
22.2 Preventing Mold-Related Nondisclosure Claims
22.3 Court Rejects Employees' Claim against Employer for Fraudulent Concealment of Mold
22.4 Incident Reports are Held to be privileged
22.5 Homeowners Policy Unambiguously Excluded Coverage for Mold

23.0 Site Safety
23.1 Store Owner not Liable for Injuries Sustained by HVAC Contractor's Employee
23.2 Engineer Had No Duty to Warn General Contractor's Employee of Danger
23.3 No Liability Under New York Labor Law for Project Owner and Lender where Worker's Injuries Attributed Solely to his Own Fault

24.0 Standard of Care
24.1 Architect Required to Review Adequacy of Engineer's Structural Report Before Proceeding with its Design Services
24.2 Summary Judgment against Plaintiff who Failed to Provide Factual Support that she could Meet Burden of Proof of Negligent Design

24.3 Design-Build Engineer Held Liable for Negligence

25.0 Surety
25.1 *De Facto* Takeover: Are a Surety's Rights Protected?
25.2 Sureties Walk a Fine Line Between Contractor Default and Claim Investigation

26.0 Time Limitations on Suits
26.1 California Decision Erodes Certainty of 10-Year Statute of Repose against Construction Defect Claims
26.2 Statue of Limitations for Negligence Instead of for Breach of Contract Requires Dismissal of Action against Architect

27.0 Warranty
27.1 Contractor Entitled to Rely Upon Government's Implied Warranty of Specifications
27.2 No Warranty of Design by Engineer

Notes

Notes

Chapter 1

Accord & Satisfaction

1.1 Accord and Satisfaction Barred Contractor Claim for Additional Compensation
1.2 Accord and Satisfaction Language Barred Contract Claim
1.3 Waiver and Release in Change Order Bars Further Recovery
1.4 Do You Really Want to Cash that Check?

1.1 Accord and Satisfaction Barred Contractor Claim for Additional Compensation (Vol. 7, No. 8)

Where a contractor timely completed a construction contract for the U.S. Army Corps of Engineers and accepted payment which included changes required by the Corps, the contractor was barred from later filing a claim for additional compensation. The claim was barred by accord and satisfaction because when signing change orders the contractor did not reserve its rights to assert any additional delay or impact. Moreover, the contractor had signed releases in certain bilateral modifications that barred its claims.

The contractor was paid the lump sum contract price plus over $600,000 in additional compensation for changes to the work that were agreed upon and authorized by bilateral modifications. After completing the work and being paid, the contractor, Jackson Construction Company (Jackson), sought to recover additional compensation under two separate claims. One claim was for early completion delay. The other was an impact claim to recover additional overhead and general administrative costs. Jackson asserted that the project was delayed by the Corps of Engineers at the start of the project causing the contractor to have to overcome a 120 day delay. This was allegedly due to the having to redesign and relocate a waterline that ran through the footprint of the building. Jackson argued the applicability of what is known as the Eichleay formula for calculating additional home office overhead. The other claim was for additional damages allegedly incurred due to cumulative impact of numerous changes required by the Corps during performance.

The U.S. Claims Court found in favor of the Corps and denied the claims due to (1) accord and satisfaction and (2) on the merits, the contractor failed to prove its case. As explained by the court, "The doctrine of accord and satisfaction is an absolute defense that terminates any previous right that a party may have had to assert a claim of the same subject matter." Further explaining this, the court stated: "An 'accord' is a contract under which both parties agree that one party will render additional or alternative performance in order to settle an existing claim made by the other party, and 'satisfaction' is the actual performance of the accord. A party asserting an accord and satisfaction defense must establish four elements: (1) proper subject matter; (2) competent parties; (3) a meeting of the minds; and (4) consideration."

The issue concerning the delay was that the contractor asserted that although it completed the contract exactly on time, it could have completed its work much earlier if it had not been for delays due to the waterline location that were allegedly caused by the Corps. In reviewing these issues, the court pointed out that Jackson chose to move the waterline and that the Corps approved Jackson's plan with the work to "be performed at no additional cost to the Government." Eventually, the Corps paid Jackson on a claim for the waterline relocation in the amount of $15,212 on a total claim of $18,212. The modification, along with all the other bilateral modifications executed by the parties included the following stipulation regarding schedule delays and impact changes: "The contract period of performance remains the same. It is further understood and agreed that this adjustment constitutes compensation in full on behalf of the contractor and his subcontractors and suppliers for all costs and markup directly or indirectly, including extended overhead, attributable to the change order, for all delays related thereto, and for performance of the change within the time frame stated."

The Corps agued that Jackson was barred by this release language from obtaining any further damages. The court agreed that Jackson did not make a timely reservation of rights to assert additional damages for the problems associated with the work, either in its original claim or at any time prior to the execution of the bilateral modification. It was not until after the contract was completed that Jackson first attempted to make an additional delay claim. The same was true of Jackson's cumulative impact for the changes required by the Corps. The court found that "Jackson never attempted to reserve its rights to assert a cumulative impact claim at any time during the work."

Jackson argued that there had been no accord and satisfaction because the release language was ineffective for the following reasons: (1) contractor says it made a written reservation of rights to assert an impact claim at a later date; (2) the modifications were ambiguous; (3) the modifications were the result of misrepresentations by the Corps; and (4) the modifications were executed under duress. The court found none of these arguments were supported by the evidence presented at trial. The court found Jackson did not explicitly reserve its rights to assert a delay or impact claim at a later date. Even if Jackson did attempt to reserve its rights, however, the court found it did not do so in a timely manner. In fact, the court stated that the delay and impact claim appeared to be an after thought. According to the court, "It does not appear that Jackson had even considered submitting a delay claim at that time. Jackson did not request any delay damages when it submitted its [original] claim for the waterline location...."

Much later, when executing two contract modifications completely unrelated to the claims at issue in this particular matter, Jackson attempted to make two after-the-fact reservations of rights pertaining to the earlier matters. "The Court views Jackson's actions as an invalid attempt to revive claims that had already been extinguished by its execution of bilateral modifications." Moreover, said the court, "Jackson did not reserve its rights to assert future claims arising from any of the other specific performance problems that it presented at trial."

In reviewing the language of the release, the court found that it was neither ambiguous nor obtained by misrepresentation. And, finally, the court found that Jackson had been under no duress when it agreed to the release language. For these reasons, the court found in favor of the Government and against the contractor.

Jackson Construction Co., Inc v. The United States, 62 Fed.Cl. 84.

Comment: This case points out how important it is for contractors to exercise extreme caution when executing change orders—if they intend to reserve any potential claims for delay and impact. It is sometimes difficult when agreeing to a change order to know the full impact of the change in the event that the project owner issues multiple additional changes impacting the project and the contractor's work. If there is a chance that additional damages may arise out of a change, the contractor should strive to include some language explicitly reserving its rights. Otherwise, it may find itself in the situation described in this case.

About the author: Kent Holland is a construction lawyer with the law firm of Wickwire Gavin, P.C., in Tysons Corner, Virginia, and is risk management consultant for the environmental and design professional liability unit of Arch Insurance Group in New York. He is also publisher of ConstructionRisk.com Report.

1.2 Accord and Satisfaction Language Barred Contract Claim (Vol. 5, No. 4)

Where a statement was conspicuously contained in a letter transmitting a check for final payment to a contractor stating that the check will constitute full satisfaction of a contractor claim, a court held this to be an accord and satisfaction, barring the contractor from claiming additional monies under its contract. In *Gelles & Sons General Contracting, Inc. v. Jeffrey Stack, Inc.* (No. 012319, Fairfax County, VA 2002), the contractor ("Gelles") had performed brick laying work. Its customer ("JSI") took issue with the amount and quality of the work and disputed the amount of compensation due. The final paragraph of its letter transmitting the check in question stated, "JSI Paving and Construction stands by its final amounts as stated on the latest correspondence.... Enclosed, please find a check .... representing final payment of the contract." Gelles cashed the check. Then Gelles sued JSI for the balance that it deemed owed. In response, JSI asked the court for summary judgment on the basis of accord and satisfaction. The court granted the summary judgment for several reasons.

Under the Virginia Code, section 8.3A-311, if a person tenders a payment instrument to a claimant as full satisfaction of the claim which was unliquidated or subject to a dispute, and the claimant obtained payment of the instrument [i.e., cashed the check], the following conditions apply. "b) ... the claim is discharged if the person against whom the claim is asserted proves that the instrument or an accompanying written communication contained a conspicuous statement to the effect that the instrument was tendered as full satisfaction of the claim."

"Conspicuous," as defined in the Code, means a term or clause that a reasonable person "ought to have noticed." There is no statutory requirement that the term or clause must be displayed in a particular font or size type. Gelles argued that the language was ambiguous and would not lead a reasonable person to conclude that full satisfaction of the claim was intended. The trail court found that a reasonable person could not have considered the language of the letters to be anything other than "a drop-dead letter that says, 'This is it. This is what we're going to pay you.'" The appellate agreed that "the entire course of conduct and communication between these parties made clear that JSI offered the [amount] as the final payment that it intended to make and that JSI considered that amount to represent the proper accounting under the contract."

Comment: I have had the experience while litigating a contractor claim to have a project owner successfully persuade a judge that through a combination of a change order, payment authorization, and partial waiver and lien release, the contractor had agreed to a total accord and satisfaction of any and all claims related to any work related to the change order. This came as a surprise to the contractor who never intended by signing the documents and cashing the check to waive its entitlement to the amount that it alleged to remain contested. Despite correspondence and documentation submitted to the owner by the contractor prior to the "accord and satisfaction," demanding payment and seeking to preserve its rights, a court held that the subsequent contractor claim for the outstanding balance could be barred. The lesson learned is that language asserting a release and satisfaction must be taken most seriously. To assure that any right to a claim for additional monies is preserved, appropriate communication and action must be taken.

1.3 Waiver and Release Language in Change Order Bars Further Recovery (Vol. 6, No.7)

Contractor that entered into a number of change orders for additional payments from the project owner (U.S. Navy) was barred from claiming entitlement to additional compensation when the Government subsequently terminated it since each of the change orders contained language releasing the Navy as to the matters covered by the change orders.

In *J.C. Equipment Corp. v. Gordon England*, 360 F.3d 1311 (2004), the contractor ("J. C. Equipment") had a fixed-price contract to repair a water system for a Navy base. Work didn't proceed smoothly due to a large number of underground obstructions and pipelines being encountered. Forty-two change order modifications were executed during performance. Pursuant to the construction contract, each of these "bilaterally executed modifications ... released the Government from further payment to J.C." for the matters covered."

The Navy eventually ordered the contractor to stop work, and the Government terminated the contract based on the contractor's alleged failure to diligently prosecute the work. The contractor filed a formal claim seeking an equitable adjustment of over $2 million. When the Government contracting officer rejected most of the claim, the contractor appealed to the Armed Services Board of Contract Appeals.

The written record included 15 thousand pages. There was a hearing resulting in a 1,000 page transcript. The contractor lost on all but a few minimal claims, and appealed the matter to court.

In reviewing the facts of the matter, and the Board's decision, the appellate court concluded the waiver and release contained in the change orders was clear and unambiguous, barring contractor recovery. The waiver and release clause was as follows: "Whenever the Contractor submits a claim for equitable adjustment under any clause of the contract ..., such claim shall include all types of adjustments in the total amounts to which the clause entitles the Contractor, including but not limited to adjustments arising out of delays or disruptions, or both, caused by such change. Except as the parties may otherwise expressly agree, the Contractor shall be deemed to have waived (i) any adjustments to which it otherwise might be entitled under the clause where such claim fails to request such adjustments, and (ii) any increase in the amount of equitable adjustments additional to those requested in the claims."

The court explained that this clause meant that in seeking an equitable adjustment, the contractor must include all items for which such an adjustment could be sought. It further meant that failure to do so would constitute a waiver of claims that were not asserted and that the contractor would execute a release of such claims as part of the receipt for equitable adjustment for any change order. The contractor's failure, at the time of the change order modifications, to explicitly except the additional items that it is now claiming from the waiver and release, bars the contractor from asserting them subsequently, concludes the court.

Comment: This decision, similar to the one discussed in the September issue of the *ConstructionRisk.com Report* further demonstrates the importance of waiver and release language. Such language when contained in an executed change order can be used effectively to manage the change order process so as to avoid subsequent disputes over entitlement to additional equitable adjustment for matters covered by the change orders. Further entitlement to either (or both) time and money can be waived and released. The language doing so should be clear, unambiguous, concise, and as comprehensive as possible if the project owner desires to eliminate subsequent claims litigation.

Where clear language concerning the release is both in the contract and in the change order (as it was in this case) the goal of the project owner will be to seek summary judgment against the contractor and thereby avoid the type of expensive litigation that must have occurred here that produced a 1,000 page transcript and up to 15,000 pages of written record. Contractors desiring to preserve their rights to further recovery must exercise due care to negotiate out of change orders language that would release the owner from further time or money that the contractor may be intending to seek later in the project that arguable arise out of or are related to these same change orders. For example, if the contractor wants to preserve the right to seek delay and impact costs that are not yet known at the time of the change order but which could result if the number of change orders keeps growing, this should be specifically stated as an exception to the waiver and release language.

1.4   Do You Really Want to Cash that Check?
(Vol. 6, No. 6)

By: *M.K. Holohan, Esq*

Everyone involved in construction contracts is concerned about payment. Owners generally refuse to pay general contractors until they have secured lien releases. General contractors are hesitant to pay subcontractors until specific blocks of work have been completed. Subcontractors are concerned about getting paid at all. Given these concerns, it is easy to see why a subcontractor might be quick to accept any payment from the general contractor. But accepting partial payment can have unexpected and harsh consequences. Subcontractors should think twice before cashing a general contractor's check for partial payment. Accepting partial payment while there is an existing payment dispute — whether related to change orders, delay costs, or some other factor — may foreclose the subcontractor's right to pursue a claim for the full amount owed. If the paperwork or check itself contains language that states that the payment is meant to satisfy the claim (e.g., "paid in full," "final payment"), the subcontractor's acceptance of that partial payment may prohibit a future claim for the remaining amount owed. This concept is known as "accord and satisfaction."

Accord and satisfaction is a legal principle that modifies a contract, allowing the parties to essentially re-write the terms of the original agreement. In this context, the "accord" occurs because the parties re-write the contract – when the general contractor offers partial payment and the subcontractor accepts. The "satisfaction" occurs because by accepting partial payment, the subcontractor gives up its claim for any additional payment. Accord and satisfaction is the law in virtually every jurisdiction in the country. The concept has been formalized by a provision of the Uniform Commercial Code (UCC), a standard set of laws governing commercial transactions, and 49 states have adopted their own versions of the UCC. Each state has the power to modify the UCC provisions; therefore, it is important to check the controlling law in each jurisdiction to determine the state laws regarding "accord and satisfaction" of contract terms.

A further complication is that the jurisdiction controlling the "accord and satisfaction" agreement may be different from the jurisdiction controlling the underlying construction contract. Because the accord and satisfaction agreement is considered to be separate from the original contract, jurisdiction agreements (known as "choice of law" clauses) in the original contract may not be applicable to the accord and satisfaction agreement. For example, if the original contract said that disputes would be decided under Virginia law, but the subcontractor cashed the partial payment check in Illinois, a Virginia court will apply Illinois law to determine whether accord and satisfaction has occurred.

Generally, to establish accord and satisfaction there are three requirements. First, there must be a bona fide claim dispute, i.e., there must be some real reason the prime contractor is disputing payment. An unjustified refusal to pay the subcontractor does not qualify as a bona fide dispute. Second, the prime contractor must have made a good faith offer to pay the claim in full. Third, the check or accompanying paperwork must contain a conspicuous statement that the payment is intended to be full satisfaction of the claim. "Conspicuous" does not necessarily mean that the language must appear in bold type or a specific color or font. Courts have found a "conspicuous statement" where the memo line of the check read "paid in full," and a letter accompanying the check clearly stated that payment was intended as "full and final payment." So, a notation on the memo of the check, or paperwork accompanying the check, will satisfy this requirement. Crossing out the "paid in full" language, or making a note on the check or document that the amount is disputed will have no effect. If the subcontractor has notice that the payment is meant to satisfy the debt, but accepts the partial payment anyway, it waives the right to pursue the disputed amount.

Although the subcontractor's claim is presumed to be satisfied when the check is cashed, this presumption may be challenged by showing that the "conspicuous notice" language was not clear enough for a reasonable person to understand that payment was for final settlement of the claim. So what standards are used to determine what a reasonable person would understand about the payment terms? The Supreme Court of Virginia recently provided guidance in *Gelles & Sons General Contracting, Inc. v. Jeffrey Stack, Inc.*

In *Gelles*, a construction company (JSI) contracted with a contractor (Gelles) for paving work. JSI received a $91,932 invoice from Gelles, and made partial payment of $70,486. When Gelles later invoiced JSI for a $26,175 balance, JSI sent an accounting of expenses incurred to properly complete Gelles' work, with its estimate for a reduced amount owed to Gelles of $13,580. Gelles disputed the accounting, and JSI responded with a letter outlining defects in Gelles' work, and the following statement: "JSI . . . stands by its final amounts as stated on the latest correspondence, dated December 8, 2000. Enclosed please find a check in the amount of $13,580.00 representing final payment on the contract." Gelles cashed the check.

Gelles filed a motion for judgment against JSI and its bonding company for $26,000 plus interest. JSI argued that Gelles' claim was barred by Virginia's accord and satisfaction law. Gelles argued that the language was not conspicuous, and that it would not inform a reasonable person that the payment was meant to be full satisfaction of the claim. Gelles asked the court to create exact standards for identifying specific conspicuous language. The court refused, reasoning that the statute's language simply required a statement "to the effect" that payment would satisfy the claim; no magic words are required to create an accord and satisfaction. The court emphasized that the merits of each case will determine whether a reasonable person should have understood that the payment was for final resolution of the claim. The court pointed to two letters from JSI to Gelles that clearly expressed JSI's intent that the payment be final resolution of Gelles' claim, and held that Gelles' claim was barred by accord and satisfaction.

The lesson for subcontractors is clear: if you have a payment dispute with the general contractor, read carefully, and do not cash a check for partial payment if there is any type of statement about full payment, or satisfaction of claims. However, even if you do cash the check, you may have a second chance to preserve your claim. In some states you will have a certain amount of time, generally 90 days, to return the payment to the prime contractor with notice that you are still pursuing your claim for the full amount owed. Then, you are free to pursue your full claim against the general contractor without being subject to an accord and satisfaction defense.

The lesson for prime contractors may be: don't wait to pay subcontractors until final resolution of all the job issues. By offering the subcontractor partial payment, plus conspicuous notice that the payment is intended as final settlement of the claim, you may be able to protect yourself from any future actions related to the claim. A subcontractor who accepts such payment will waive the right to pursue other, and potentially larger, claims against you.

About the Author: M.K. Holohan formerly worked as an associate in Wickwire Gavin's Vienna, Virginia office.

Chapter 2

Americans with Disabilities Act

2.1 Design Professionals Not Subject to Liability under Title III of the ADA and Washington's Law against Discrimination (Vol. 6, No. 1)

By: Bruce Marvin and Beth Andrus

The Federal District Court for the Western District of Washington has recently held that Title III of the Americans with Disabilities Act (ADA) and Washington's Law Against Discrimination (WLAD) do not provide a statutory basis for direct or third party claims against design professionals.

Generally, Title III of the ADA prohibits discrimination against the disabled with regard to access to places of public accommodation. It provides: "No individual shall be discriminated against on the basis of disability in the full and equal enjoyment of goods, services, facilities, privileges, or accommodations of any place of public accommodation by any person who owns, leases (or leases to), or operates a place of public accommodation. 42 U.S.C. 12182(a).

Types of discrimination expressly prohibited by Title III include the design and construction of facilities that are not readily accessible and usable by individuals with disabilities. 42 U.S.C. 12183(a)(1).

Courts interpreting WLAD generally apply the same analysis used to interpret federal antidiscrimination laws. Accordingly, prohibitions against discrimination on the basis of disabilities in WLAD are subject to the same analysis applicable to ADA .

In *Marshall v. Cafaro, Co.*, a disability rights group sued a mall owner for violation of Title III. In response to this lawsuit, the Owner filed third party complaints against various design professionals and contractors who had worked on expansion of the mall over the previous 10 years. Among other things, the Owner asserted third party claims against one of the mall's architects, including breach of contract, professional negligence, and statutory and common law indemnification and contribution.

With regard to the statutory claims, the Owner alleged that the Architect's design failed to conform to ADA requirements and, therefore, the Architect should be held liable for any damages the Owner might incur as a result of the plaintiff's ADA and WLAD claims.

The Architect responded with a general denial of all liability based on the fact that the Architect's design satisfied all ADA requirements in effect at the time of construction. When the Owner refused to voluntarily dismiss its claims, the Architet filed a motion for summary judgment, arguing that the Owner had failed to state a cognizable claim under Title III of the ADA or WLAD. The Architect also argued that the Owner's contract, tort, indemnification and contribution claims were barred by the applicable statutes of limitation and by Washington 's six year statute of repose for construction-related claims.

The District Court granted summary judgment and dismissed the Architect from the case. Rejecting the Owner's Title III contribution claim, the Court cited to *Lonberg v. Sanborn*, 259 F.3d 1029, am. 271 F.2d 953 ((9 th Cir. 2001), in which the Ninth Circuit refused to extend Title III liability to designers and contractors responsible for construction of building that failed to conform with the ADA. In doing so, the Ninth Circuit found that the "general" rule set forth in 42 U.S.C. 12182(a), which specifically limited Title III liability to persons "who own [], lease [] ...,, or operate [] a place of public accommodation" clearly established Congress's intent to restrict ADA liability to owners, lessors and operators of public facilities.

Although noting that Lonberg did not specifically address contribution claims, the District Court found that the Ninth Circuit's holding that architects were not within the scope of parties subject to Title III liability was persuasive. Accordingly, it concluded that the Owner had failed to meet its burden of establishing that there was an implied right of contribution available against architects under Title III. The court, therefore, dismissed the claim.

Because the WLAD is subject to the same legal analysis as the ADA , the District Court held that the Owner's contribution claim under WLAD failed for the same reasons that its ADA claim failed.

Conclusion. All design professionals have a professional and ethical obligation to design structures that comply with the applicable laws and building codes, including the ADA . While Marshall and Lonberg stand for the proposition that Title III does not provide a statutory cause of action against design professionals, failure to comply with the ADA may still result in liability for malpractice or breach of contract (claims the Marshall court did not have to address because they were barred by the applicable statutes of limitation and the six year statute of repose).

About the authors: Bruce Marvin and Beth Andrus are attorneys with the law firm of Skellenger Bender, P.S., 1301 Fifth Ave., Suite 3401, Seattle, WA 98101. They may be reached at bmarvin@skellengerbender.com and bandrus@skellengerbender.com.

Chapter 3

Changes

3.1 Managing Contract Changes (Vol. 6, No. 3)

By: *J. Gerard Boyle*

Shortly after work commences on design-bid-build projects all over this country, it can be predicted with reasonable certainty that the Owner and/or Consultant will confidently announce to the General Contractor (GC) and its subtrades: *"There will be no contract changes on this project"*. This they may say after 10 addenda have been issued during a three-week tender period, and despite the fact that they themselves have never been on a project without contract changes (and often a great many). If they actually believe this dictum can be observed, they suffer from the sort of wishful thinking that had Francis Fukuyama announcing the end of history. And yet while a GC may shake its head at such idle pronouncements, knowing full well that changes (along with the delay, disruption and consequential effects they frequently cause) are inevitable, the GC often conducts itself as though totally unprepared to deal with them.

The following discussion suggests that the GC can manage changes in such a way that its exposure to claims is minimized and its profitability is preserved, while at the same time avoiding, or a least minimizing, the deterioration of relations with the Owner and Consultants prevalent where change management is wanting. Effective change management is conducive to the health of the project; an interest shared by all project participants, each of whom has a part to play in ameliorating the potentially consumptive effects of changes to the work.

Changing the Climate: The Role of the General Contractor

The GC must take the lead in getting the project participants to realize that all parties have an interest in the early identification and resolution of changes because: a) failure to do so will lead a project into a morass of conflict and contentious claims; and b) nobody else is going to do it. The GC's project administrator must often do this in a climate of mutual suspicion that the project participants carry as baggage from past disasters. For their part, Owners are convinced that Contractors have a vast apparatus dedicated during the bidding stage to change discovery, so that they can exact their ransom on the job absent competitive pressures. GCs, confronted too often by Consultants who defend deficient documents against reasonable claims for extra, become totally skeptical about the Consultant's objectivity.

The truth that Owners should know is that GCs barely have time to assemble a bid, let alone to attempt the sort of predatory 'change discovery' strategy posited by the 'Contractor conspiracy' theory. GCs gain perspective if they realize that the Consultant is often deprived of the resources required to produce more complete documents, and may have an Owner muttering 'errors and omissions', each time he presents a change order for signing. Contending with, and overcoming, the prevailing prejudices of the project participants is an extremely important first step in creating an environment conducive to the resolution of changes.

There is a compelling need for leadership in this regard, and if the other project participants are disinclined, it is in the interest of the GC to assume the leadership role. Discussions about leadership often appear to float up in a balloon filled with the heady air of idealized assumptions about a utopian project where the players, once enlightened, are readily disposed to rise above their narrow self-interest. The type of leadership called for here is not of this sort: it is pragmatic, not Pollyannaish; it is earned, not imparted.

The leadership recommended here can mean dollars in a GC's pocket and avoidance of unnecessary exposure to Owner and subcontractor claims. First the GC demonstrates that if he has the information he can build according to schedule. Next he shows that he has control of his subs, and that if they do not perform he will take swift action. He knows where he has been and where he is going, by way of attention to scheduling. He controls the project documentation, especially the flow of changes. He has a thorough understanding of his contractual[1] rights, but conveys this without intimidation. He is so confident in his leadership position that he can manage forcefully, but judiciously, utilizing effective conflict-mitigating strategies such as interest-based, rather than position-based, approaches.

Absent this type of leadership, scenarios such as the following may be expected.

Mismanaging Changes: A Typical Case

During the third month of a scheduled nine-month project, the Owner directs the Architect to prepare a Contemplated Change Order (CCO) that will involve changes to the Architectural, Civil, Mechanical and Electrical scope of work. Two weeks pass before the CCO is issued, and the work involves a change to one area of the building from the currently required classroom arrangement to an office space. The layout of the affected area, including wall and ceiling layouts, will be completely revised. Light fixture and diffuser locations and types will be changed; re-ordering will be required.

Some M&E rough-in and drywall stud installation has already been completed. The GC instructs the trades to stop work in this area in accordance with the Owner's verbal instruction – the Owner told him unofficially the change will definitely be implemented because it has been directed from 'higher-ups'.

This is a renovation project and by this time there have been numerous other (especially structural) changes: mostly Owner-directed (by a CCO), but also a growing number of changes for which no CCO has been issued. Given the upsurge in price requests, and his growing commitments on the other two jobs he is running, the GC is falling behind in delivering the CCOs to the affected subtrades, and in this case almost a week goes by before the change is faxed out. It doesn't help him that some trades, especially the painting subtrade are consistently slow to price.

The GC is actually not aware that the Mechanical scope definition for this change is incomplete and that the Mechanical subtrade has been bypassing the GC, pushing the Mechanical Consultant for information so that he can quote the work. The structural scope of work is not accurately depicted in the CCO: major alterations to the approach have been worked out on site. The GC instructs his steel subcontractor to ignore the obsolete CCO and quote the work as it will be constructed.

About four weeks after receiving the change the GC has assembled about ten quotations which he hands to the Architect at the next regular site meeting. During the site meeting, the GC becomes aware by way of the complaints of his own superintendent and some of the trades that this change (and others) is now seriously affecting the progress of the work. The architect looks over to the Owner and observes that he has not even received pricing for the change. The GC quickly acknowledges and apologizes for the delay in pricing, but explains that the quotation is included in the package he has before him. "In the interest of schedule", the GC asks the architect, "can we have an immediate opinion on the price."

The Architect objects, advising he will need more time to review the quote, and then notices that the painting price does not include a break down for the labour and material portion of the work. He chastises the GC: "This is not the first time that changes are being submitted without adequate detail." The GC almost replies that the Architect has not yet responded to a single quotation but, feeling himself on shaky ground in this case, he decides that now is not the time to get into an argument.

The Mechanical Consultant, who only visits the site every other week is not available to comment on the Mechanical portion of the quote. Although the GC has succeeded (at his own risk) in convincing the steel subcontractor to proceed with the structural work in the interest of not delaying the job, the Mechanical sub will not proceed without a go-ahead. Another week passes by before the Mechanical Consultant finally sends his approval recommendation to the architect. However, the architect is convinced the Mechanical Contractor is 'gouging', and recommends to the Owner that the quotation not be accepted.

Instead, a Change Directive is issued, materials are ordered and the work is performed. Two months later the GC submits a fully substantiated quotation complete with time sheet signed by his superintendent, delivery slips and invoices for all items. At the next site meeting the Mechanical subtrade complains that he has not yet invoiced for any of the work (note: the GC will not permit billing for a change without a corresponding Change Order) that has been performed and insists on immediate issuance of a Change Order. The architect complains that the substantiated labour hours "seem very high".

The Mechanical subcontractor furiously objects, pointing out that all of his time has been duly signed by the GC's site superintendent. The Architect cuts off the discussion, concluding that since the amount of the quotation has greatly exceeded the original quote he will simply recommend acceptance of the original quote amount. The GC bangs the table and announces that no more changes will be performed without prior approval, and the meeting abruptly ends. "And forget about your schedule", he says, to which the Architect responds that he has not seen a single schedule update.

Later, the GC's project administrator is brought before his own management who at first assail him for an architect's letter accusing the GC of not pricing and managing the job changes efficiently. They soften their position once he explains that he is being inundated with changes on the job and reminds them that this after all is only one of three jobs he is looking after. Just as they all agree on a "hard ball, letter of-the-law approach" to deal with future changes, a fax is received from the Mechanical Contractor wherein he asserts that he is being delayed in the progress of the work and is being adversely affected by late payment on extras.

Identifying the problems and finding solutions

Variations on the above scenario play out on construction projects with disturbing regularity. It should first be understood that although there is plenty of blame to go around in this deteriorating situation, it is clearly the case that the GC is not playing the leading role in the resolution of changes and this is to the great detriment of his firm and the overall health of the project. It has been said that if the facts are on your side, you should hammer the facts, if they aren't, hammer the table. This GC should chastise himself before berating the Consultants, he is really banging the table because he has needlessly lost control of the project.

To begin with, if this GC actually knows his contract, he has decided to ignore provisions that would otherwise have afforded protection to his position and that of his subcontractors. The first thing that should have occurred to the GC is to consider what type of change he was being asked to perform. This change is not a straightforward addition, or 'extra', as for example when a diffuser is added to a room sufficiently in advance of the planned schedule so as not to require re-work or cause interruptions. In the case of this particular change, re-work and re-ordering will be required, and delay and disruption will result. Depending on the circumstances, such a change may fall into the category of a 'scope' change and, if it does, the GC has a choice: he may have the right to refuse to perform the work.

The contract clearly allows the Owner to "make changes in the Work ... by Change Order or Change Directive"[2]. However, the Owner **and GC** must "agree to the adjustments in contract price and contract time"[3]. Where there is no agreement, and the Owner requires the GC to proceed, the Change Directive is to be used.[4] But a Change Directive may only deal with work "within the general scope" of the contract documents.[5] Therefore, in the case of a change outside of the scope of the contract, if the GC decides that it does not want to perform the work, it could arguably simply resort to its right under the contract to refuse such work.[6]

Having said this, it must be stressed that in this context the CCDC 2 – 1994 contract does not provide a definition for a 'scope' change, nor does there appear to be a consensus in the industry on what actually constitutes a general scope change. Consequently a decision to refuse to do the work should only be considered after legal advice. The consequence of incorrectly determining work to be outside "the general scope of work" could be severe and this is likely a rare occurrence.

The next important point to be made is that the verbal instruction the Owner gave to the GC is insufficient. A stop work order, being a change to the work, should be issued as a

Change Directive[7]. If the Owner had ultimately decided not to proceed with the change, the GC would have experienced delay, but by having nothing in writing, he may have no protection under the contract.

Having apparently ignored any consideration of whether he is arguably contractually obliged to perform this change, the GC then proceeded with some elements of the work without either a Change Order or a Change Directive. Of course the wording of the contract is very clear in this regard: *"The Contractor shall not perform a change in the work without a Change Order or Change Directive."[8]* In fact, the GC did not proceed out of ignorance of this very clear stipulation; he did so knowingly, and was already performing other changed work without a change order or a change directive.

There were several factors at work in his 'reasoning': he had a level of comfort and trust with this Owner and so felt reasonably sure that the changes would eventually be approved; he did not want the schedule to suffer by delaying work he was confident would be required; he thought his willingness to proceed with changes prior to a Change Order or Change Directive would foster good will and; finally, he was reluctant to demand strict adherence to contractual procedures on changes because he felt vulnerable – he was not pricing in a timely fashion and was concerned this deficiency would become very apparent by this approach. In fact, he has fallen, by his own neglect, into the trap of proceeding with work prior to pricing, and without a Change Directive that would define the method of evaluation.

Absent the Change Directive, the Consultant, if he does not simply refuse to acknowledge the change, will often take the position that the work the GC is performing is to be evaluated as a lump sum quotation, in which case actual job site conditions may be ignored in favour of an 'objective' estimate.

As the GC sees it, he is taking a calculated risk, but such risk is really unnecessary and ill advised. Assuming that the number of changes that are issued do not become so overwhelming that it can be reasonably asserted that existing resources can no longer be expected to deal with them[9], and furthermore that the price request document is adequately detailed, then a failure of the GC to produce pricing in a timely fashion, is really an inexcusable failure because it fatally weakens his ability to demand accountability from the other project participants.

Once the GC makes a management decision to devote as much time as it takes to keep on top of pricing, it must enforce the same requirement on all of his subtrades. The GC's subcontract with the trades should include a clause stipulating a strict time frame for subtrade pricing, after which it reserves the right to conclude no adjustment to the subtrade's price. The GC's log of contract changes which he creates to record every extra cost item he identifies, should become the document of reference for change status and should be reviewed at every site meeting.

The CCOs being issued on this project are often incomplete and incorrect, and this is undoubtedly slowing pricing and affecting the quality of the submissions. The Mechanical subtrade is chasing the Consultant for a full scope, and not making an issue of it because he wants to preserve his good relations with the Consultant. Furthermore, the CCOs are not being issued quickly enough to keep up with the changes on the job. This will not change until the Owner and Consultants are taken to task. The Request for Information (RFI) document should be used to record deficiencies in the price request documents.

As evidence accumulates of consistent failure to produce adequate information to price, the GC should request in writing that the change(s) in question be issued as a Change

Directive in order to mitigate potential or actual delay to the project.

A GC has recourse by the contract to deal with the inadequacies of resources on the Owner's side and vice versa. If, as in this case, a Mechanical Consultant is not available to visit the site in order to fully evaluate some aspect of the quote, and it is delaying the process, request a Change Directive.

After all the time wasted in this process, albeit some of it due to the Contractor, the Consultant then attempts to 'negotiate' the amount of the Change Directive. The contract makes clear that the earlier-submitted lump sum price is superseded by the time and material approach (unless both parties agree otherwise)[10], the Consultant must evaluate the substantiated quote for the work performed on a time and material basis in accordance with GC 6.3.4 and on its merits.

Labour hours in particular should be signed on a daily basis, preferably by an authorized agent of the Owner or Consultant. Once this is done, there is still room for the Consultant's reasonable questions with respect to particulars of the quotes, but "seems high" does not constitute a reasonable review of the Contractor's detailed quotation. Accumulated hours should be presented to the Owner/Consultant at least once a week and the approved amount billed each month[11].

Changes that the Owner/Consultants will not Acknowledge

The foregoing example dealt with a change that was introduced by the Owner, but another significant challenge for a GC is to defend the reasonable claims for extra's that the Owner/Consultant refuse to fairly consider. The GC must be fair and reasonable, so that he will engender trust, but firm and unrelenting in demanding equitable compensation for extra work according to some fundamental principles supported by the contract, the common law and, sometimes, common sense.

The most important principle is that on a design-bid-build project, the GC may expect that the bid documents including plans, specifications and addenda convey the scope in a clear and comprehensible fashion. The GC is not a designer unless the contract specifically states otherwise (as is the case, for example, in GC 3.3 Temporary Supports where the GC must hire a structural engineer to produce a design)[12]. The GC can construct only to the extent that the design and Contract Documents permit such performance.

Experienced Contractors are familiar with lines of argument that Owners and Consultants have developed to try to contend with their own vulnerability in this regard. An attempt may be made, for example, to use the pre-bid site visit as a substitute for a documented scope of work, even though the contract documents, by themselves, should convey the scope of the work, and should not require the elaboration of site interpretations to produce a complete picture of the scope of work.

Information that is necessary, but not sufficient, is sometimes offered as a complete scope (as when for example fire extinguishers are mentioned in a specification but no quantity is given and none are shown on the drawings). Often what a Consultant calls 'coordination of the work' which is the responsibility of the GC is really 'coordination of the design'[13], for which the Consultant is responsible.

It should also be remembered that once these approaches are exposed for what they are, the GC still has to construct, and that without his prodding and contractibility input the information he requires to build will probably not be provided quickly enough. Even when the Owner/Consultant are clearly at fault, the GC can never be the indifferent bystander, he must not only be a part of the solution, he must drive the solution because if he does not, the potential damages to which all are exposed, may be magnified.

Changes are so dangerous for GCs because the delay, disruption and/or impact effects they often cause, while potentially claimable for the Contractor may instead become, if not managed properly in accordance with the contract, a situation of significant exposure to claims. The contract allows that every change should be considered with respect to the potential delay it may cause[14].

Consistent with the contract, The Canadian Construction Association recommends that GCs add a line to their quotations for "schedule acceleration/extension" as well as "impact" in their model Change Order Quotation form[15]. It goes on to suggest the inclusion of the period of days (addition or deletion) of schedule effect and contains wording reserving the right of the Contractor to assess impact of the change at a later date if such impact cannot be assessed at the time.

Why then do GCs so often fail to include for the time impact effects of a given change? Some GCs will include the exculpatory language regarding impact, but yield to the objections of the Consultant. Others will include the suggested allocation of schedule days affected by the change, but more often GCs do not directly address schedule at all. Why would a knowledgeable Contractor not attempt to assign a time effect when the contract clearly affords this right?

Sometimes it is actually not possible, in other cases, Contractors are actually concerned that in so doing, they wave their right to future claims for the accumulated affect of all changes. But in most cases it is simply because the GC has failed to update the CPM schedule, which is the contractually mandated instrument of time effects, on a regular basis.

Most contracts require the GC to update the critical path schedule on a monthly basis[16]. Moreover, written notice of delay must be given by the Contractor to the Consultant within 10 working days after the commencement of the delay[17]. The GC who fails to provide clear and timely notice may not only waive his contractual right to compensation, but in failing to assert his own rights he may leave himself exposed to claims by subs and the Owner.

The GC must understand how changes are affecting the schedule. On changes where there is a clear-cut effect on the original program, it is preferred to perform a 'snapshot' analysis[18]. However, many GCs are either not trained in the techniques of such analysis or, if they are, fail to allocate their time to this task. The great value of such a demonstration is to introduce into the contemporaneous job record a document that will record agreement at the time with respect to relevant facts pertaining to the as-built status of the job and schedule logic. If not so recorded, such disputes begin with efforts to negotiate agreement on fundamental facts that might otherwise have been established. Delay and disruption claims should be governed by the contract.

It is in the GC's interest to devote sufficient resources to realize this objective. It is entirely possible, but rarely accomplished, to settle changes and delay and impact claims during the life of the project and without acrimony or resort to legal remedies. To be sure this will depend to a large degree on the reasonableness of the project participants, but critically important is that the GC is in control of documentation (especially changes), monitors schedule, knows his contract, and exercises leadership in contending with changes and the delay and consequential effects they so often bring to a project.

Early Discovery of Changes
It was suggested above that GCs, owing to time and resource constraints, do not have time during the tender

period to perform a 'change discovery' examination of the bid documents. After award however, it is of great benefit to the Contractor to work with its subcontractors to identify changes at an early stage (say within the first month), and to then have the Consultants document the required design change so that the delay effect can be minimized. It is even suggested that the Owner and Consultant be asked by the GC to participate in this exercise, although they may not be able to see that it is in their own interest to settle the scope early. Such an early approach by the GC may, if nothing else, succeed in moving the Owner/Consultant away from the 'Ostrich' mentality on changes, so that the reality can be dealt with.

It should be understood that not all changes can be discovered at an early stage. For example, the fact that the Architect's layout for the Mechanical room conflicts with Mechanical design, may not be discoverable even by careful review of working documents, and so will only be discovered once work is sufficiently advanced. Changes of this sort may be symptomatic of a pervasive problem of incomplete, absent and/or conflicting design. In such cases, the pro-active approaches advanced here are of limited utility, and the best a GC can do is to adopt a 'damage control' posture.

There are, however, changes that can and should be identified by the GC at an early stage of the project. Included in this category are changes of the sort found on renovations projects where, for example, the Consultants rely upon the original project 'as-built' drawings, later found to be incorrect, as a layout template. In this situation the GC may find that rooms indicated in the bid documents do not even exist, wall locations may be incorrectly drawn, ceilings identified in schedules as drywall may be plaster; all involving changes to the original plan which may prove significant. Other examples of 'discoverable' changes resulting from an inadequate 'survey' of the building by the designers are as follows: the existing masonry walls require extensive repair not indicated in the documents; or the window opening is not large enough for the new louvre it is to receive, requiring a new structural steel header.

Critical to the success of this early 'change discovery' approach is that the GC's supervisor and the subcontractors understand that among the priorities during the first weeks of the project must be a wholesale site investigation and a thoroughgoing document review. Moreover, shop drawing submittals, which introduce an essential, additional level of information that may reveal design problems, must be expedited. Of course if, as is often the case, the GC is dilatory in the award of subcontracts, an essential participant in the early change discovery approach is not available, and the process cannot really get started in earnest until awards are finalized.

Conclusion

Many GCs today see themselves as 'brokers'. This thinking is encouraged by economic considerations. If project delivery for the GC only involves limited oversight of the work that others (subcontractors) will perform, overhead costs can be minimized. These days a GC 'passes through' most of the actual construction work as well as the provisions of the 'prime' contract to its subcontractors. What cannot be 'passed through', and where the 'brokerage' model breaks down, is the requirement for effective management of the project in the volatile atmosphere of contract changes, and the leadership this entails. It is hoped that this article will cause not only GCs, but Owners and Consultants as well, to realize that the success of the project requires a shared commitment to the early identification and timely resolution of changes and, equally important, the allocation of sufficient resources to accomplish the task.

About the Author: J. Gerard Boyle is Claims Analyst and Sr. Project Manager with Revay and Associated Limited, a Canadian firm with a national practice of construction consultants and claims specialists, assisting owners and contractors on projects. He may be reached in the Montreal office at 4333 Ste. Catherine Sat., West, Suite 500, Montreal, Quebec H3Z 1P9; (514) 932-2188; montreal@revay.com; http://www.revay.com. This article was originally published in The Revay Report, Vol. 23, No. 3 and is re-published by ConstructionRisk.com Report with permission.

Editor's Note: This article, although addressing the question of contract changes in the context of specific Canadian contract language has valuable insights with general application to contracts in the United States and elsewhere. With the number of subscribers to our Report that are based in Canada and countries other than the United States I believe it appropriate to include this article for our readers.

---

[1] The Standard Construction Document – CCDC 2 – 1994, Stipulated price contract, created by the Canadian construction documents committee, is referred to throughout this document.
[2] CCDC 2 – 1994, GC 6.1.1
[3] CCDC 2 – 1994, GC 6.2.1
[4] CCDC 2 – 1994, GC 6.3.1
[5] CCDC 2 – 1994, Definitions 18
[6] Unlike the CCDC 2 – 1994 document, The CCDC 2 – 1982 Stipulated Price Contract provided a definition for "Changes in the Work", and restricted changes to those "within the general scope of the contract".
[7] CCDC 20 1994, GC 6.2, p. 24
[8] CCDC 2 – 1994 GC 6.1.2
[9] At some point, the extent of changes become 'unmanageable' and beyond what the Contractor could reasonably have expected, but what is this threshold? A study by the Building Research Board National Research Council found that changes in the range of 6 – 10 percent would be expected. C. Leonard's study concluded that a correlation

between productivity and CO hours begins to appear when changes exceed 10 per cent of base contract hours. None of this is conclusive, and it is important to consider whether the changes occur early or late in a project as well as whether or not they were evenly distributed over the project period. See "Calculating Lost Labour Productivity", William Swartzkopf, 4.3, for a discussion of reasonable expectation of extras on a project.
[10] CCDC 2 – 1994 GC 6.3.7
[11] CCDC 2 – 1994 GC 6.3.5
[12] CCDC 2 – 1994 GC 3.3.1
[13] Complete Contracting: A-Z Guide to Controlling Projects, A. Civitello p. 97
[14] CCDC 2 –1994 GC 6.2.1 and 6.2.2.
[15] CCA Doc. 16 1992: Guidelines For Determining The Costs Associated With Performing Changes In The Work.
[16] CCDC 2 – 1994 GC 3.5.1.2
[17] CCDC 2 – 1994 GC 6.5.4
[18] Construction Claims Causes and Options, S. G. Revay, p. 107

Chapter 4

Contractor Claims

4.1   Contractor Suit Dismissed for Failure to Follow Claim Procedures of Contract
4.2   Failure to Request Change Order Bars Contractor Recovery for Excess Units
4.3   Multi-Million Dollar Claim Invalidated by Court Due to Contractor's Failure to Give Timely Notice
4.4   Absent Contractual Allocation of Delay Risk, Subcontractor Cannot Recover Damages from General Contractor who was not Responsible for their Occurrence
4.5   Equitable Adjustment Allowed for Deductive Change Despite Contractor's Unbalanced Bid
4.6   Proving Constructive Acceleration Claims Can be Difficult for Contractors
4.7   When to Stop Work for Non-payment
4.8   Differing Site Conditions, Defective Specifications: One Coin, Two Sides

4.1   Contractor Suit Dismissed for Failure to Follow Claim Procedures of Contract (Vol. 6, No. 3)

A homebuilder's lawsuit against a city and its architectural firm for refusing to grant change orders for additional costs was rejected by a court because the contractor failed to comply with a contractual requirement that it give written notice to the architect of its breach of contract claims against the city before initiating litigation.

In the case of *Cameo Homes v. Kraus-Andersen Construction Company and City of East Grant Forks*, (No. 04-1200, U.S. 8th Circuit), the Court of Appeals affirmed a summary judgment granted in favor of the City and architect by the federal district court. Federal District Court . Cameo Homes (hereinafter, "Cameo" or "Contractor") entered into four contracts with the city to do concrete work on four projects being managed by Kraus-Andersen, the city's construction manager (CM). The contracts incorporated the general conditions of the American Institute of Architects (AIA) contract provisions, including a provision stating that the contract could only be modified through written agreements such as "change orders."

In the section addressing change orders, the contract provided that change orders requested by the contractor were then to be prepared by the construction manager stating the additional work to be performed, the deadline and the additional compensation, if any. No change order would be effective unless singed by the CM, the architect, and the contractor. No oral modification of contract terms was permissible.

In addition to the change order process, the contract addressed a separate category called "claims." The difference in a change order and a claim is that change orders modify the terms of a contract while claims seek relief owed "as a matter of right" under the existing terms of the contract. Written notice of any claim was required to be submitted to the architect by the contractor within 21 days of an event or the discovery of an event giving rise to the demand. As a condition precedent to any litigation, the contractor was required to first obtain an architect's decision of a claim.

In this case, Cameo, the contractor, submitted several change orders which were denied. Cameo performed the work demanded without obtaining change orders and subsequently submitted change orders for its extra costs. These change orders too were denied. Cameo then filed suit against both the city and the CM, alleging breach of contract by the city, negligence by both the city and the CM as well as defamation, fraud, RICO violations, intentional and negligent interference with prospective business advantage, intentional interference with contractual relations, extortion, civil conspiracy, and violations of the Davis-Bacon Act.

After the defendants had the case removed to the federal district court, they succeeded in having the court grant their motions for summary judgment and dismiss all the contractor's claims. On appeal, Cameo argued that the district court erred in determining that it had failed to give notice of its claims to the architect. In particular, the contractor argued that the process for submitting a claim "is referred to as the 'change order process.'" In practice, the contractor claims that the parties amended the claims process by allowing change order requests to be submitted and approved after contested work had been performed. Cameo therefore argued that its submission of change order requests satisfied the requirement that written notice of claims be given to the architect before litigation.

The appellate court concluded that although the contractor submitted some evidence suggesting the parties developed an alternative practice for the approval of change orders, it had not proved that the parties understood change order requests to be equivalent to submitting formal claims to the architect as required by the contract. The claims process was never modified in writing. Since the contractor failed to adhere to the contract's strict requirement to give written notice of claims to the architect, the court held the contractor was barred from filing suit in court. This applied not only to breach of contract claims but also to the alleged negligence-based claims against the architect.

The court concluded that the contractor's disagreement with the architect over verification of concrete placements was subject to the same contractual claims process rather than a negligence action because the architect acted within the scope of the contracts and incurred no corresponding obligation directly to the contractor. For these reasons, the appellate court affirmed the summary judgment against the contractor.

Comment: This case once again demonstrates the importance of knowing and following the contractual requirements concerning the change order process and claims process. There are numerous reported decisions in courts throughout the country upholding the contract terms to bar contractor claims that failed to meet the time requirements or notice requirements specified. As explained by this court, when a contract states on its face that it can only be amended in writing, it means what it says. Even if the parties appear to informally waive the contractual requirements to permit consideration of changes or claims that do not adhere to the literal black-letter requirements of the contract, many contracts state that such an apparent waiver on one matter does not constitute a waiver of the provision going forward.

Even where no harm or apparent prejudice has been caused to the architect or project owner by a contractor's failure to adhere to the strict claims process requirements, numerous court decisions have held that the failure to comply with the contractual claims process acts as an absolute bar to filing claims in court. The basic idea is that where commercial entities have bargained for and entered into a contract, it is not proper for courts to undo or rewrite the bargain.

As part of the risk management training that I present to owners, design professionals and contracts, I emphasize the importance of knowing your contract. Know what it says with regard to time limitations for filing change order requests and claims. Know what it says concerning where to file and with whom to address the change order requests and claims. Be sure that the responsible field personnel and project managers know the requirements. And be sure they all follow the requirements.

4.2 Failure to Request Change Order Bars Contractor Recovery for Excess Units (Vol. 6, No. 7)

Contractor that performed significantly greater unit quantities of paving work than anticipated was barred from an equitable adjustment. It adequately documented the increased quantities and costs. The project owner was aware of the facts,. The contractor failed, however, to submit a change order for approval by the owner in advance of incurring the costs.

In *Seneca Valley, Inc. v. Village of Caldwell*, 808 N.E.2d 422 (Ohio, 2004), a contractor (Seneca Valley), on a waterline project, brought suit against a village seeking an equitable adjustment for increased costs associated with excess materials and costs it incurred for increased unit quantities of pavement. The contractor argued that the village was in breach of its contract by failing to pay for the work, or that in the alternative, the contractor was entitled to recover its extra costs under the theory of "quantum meruit" because the village would otherwise be unjustly enriched by all the extra work performed by the contractor.

The contract had been awarded to Seneca as the lowest bid on a lump-sum, fixed-price contract. The specifications included in the invitation for bid showed a small unit quantity of pavement replacement. Seneca asserted in the litigation that the contract was not a lump-sum contract but a unit price contract. The contractor placed approximately 1,380 cubic yards of granular aggregate and replaced 1,422 square yards of pavement at the project site. (Only ten square yards of asphalt restoration was included in the unit price estimates contained in the bid and contract).

No written change order was issued for the placement of these materials. It was not until the contractor's pay estimate was submitted to the village seeking payment for these quantities that the dispute became apparent. The project manager for the village responded to the pay request by advising that, "A quantiy increase over the bid quantity of this magnitude needed a request for a change order." In addition, the project manager noted that the material used for the pavement restoration was not the same as the bid item and that this also required a written change order to approve this change.

In considering the matter, the court stated that it was undisputed that the contractor performed work beyond the specific base units set forth in the bid. The court stated that the question to be determined was whether the contract was a unit base contract (fixed price per unit but no set number of units) or a fixed bid based contract (one set price for completion of the job as a whole). The contractor argued that the contract was a fixed bid contract and that setting unit prices within the contract was "merely a convenience for calculation of the final bid and, in the event that the specific number of units listed in the contract was exceeded, was to be used in order to more easily facilitate a written change order."

The village's Notice of Award to the contractor stated "You are hereby notified that your bid has been accepted for items in the amount of $103,040." This amount, said the court, was the total amount of the lump sum bid and the notice of award didn't anywhere mention unit prices. The General Conditions section of the contract, says the court, support the villages argument that the unit prices that were identified in the contract were "pre-negotiated unit prices for dealing with post-contractual increases."

The court quoted at length from various contract provisions concerning "Contract Price," "Field Orders," "Change Orders," "Measurements and Quantities," and "Changes in the Work." It's conclusion was that in accepting the contractor's bid, the bid was accepted based on a finite amount of money and did not refer to unit prices. The court finds: "Further, the fact that the base bid quantities may have been estimates does not eliminate the change-order requirement for work quantities in excess of the estimates. This contract unambiguously requires a written change order in advance of any additional pay items." Based on this, the court found that the village did not breach its contract by refusing to pay the extra costs demanded by the contractor.

Contractor's alternative theory for recovery (i.e., unjust enrichment for the village) was also rejected by the court.

As stated by the court, "While the outcome herein may render unfortunate results for the [contractor], it is not the function of this court, or any court, to construe an otherwise unambiguous contract in order to achieve equitable results. Finally, the court rejected the contractor's argument that the village had waived its rights under the contract by virtue of the fact that its project representatives had approved daily work in excess of the initial bid item quantities.

As explained by the court, "The contract documents in the instant case do not provide that the Village water department employee, acting as an "Observer," had authority to authorize contract quantities orally or even writing. There is no evidence of, or any claims of, an express waiver by the Village. In fact, the bid contract documents themselves negate such a claim." For these reasons, the appellate court affirmed summary judgment in favor of the village against the contractor.

4.3 Multi-Million Dollar Claim Invalidated by Court Due to Contractor's Failure to Give Timely Notice (Vol. 6, No. 1)

By: *Katz & Stone, L.L.P.*

Most construction contracts contain language obligating the contractor to submit claims for extras or changes to the owner or higher-tier contractors within a certain period of time after it incurs increased costs or delay. In addition, many contracts also require the contractor to notify the owner or higher-tier contractors of its intention to ultimately make such claims within a certain period of time after the event giving rise to the claims. As demonstrated by *MCI Constructors v. Spotsylvania County*, 2003 Va. Cir. LEXIS 115 (Spotsylvania County Cir. Ct. 2003), it is critical for contractors to understand the precise moment when the time period to take action on a claim will begin to run. Adopting an unreasonable or unrealistic view of the triggering event can have disastrous consequences.

In *MCI Constructors*, a contractor was engaged by Spotsylvania County to construct a water treatment facility. As work progressed, disputes arose between the county and the contractor, delays occurred and extra work was required. During the course of this project, the contractor submitted various change orders to the county for approval, several of which were rejected. Months after the project was scheduled to be completed, the contractor submitted a "Request for Equitable Adjustment" to the County, requesting additional payment of more than $9 million and an eight-month time extension for the changed and extra work it performed. The County disputed 93 of the 106 claims asserted within the

Request on the grounds that, among other things, the contractor failed to give timely notice of its intention to submit such claims. In response, the contractor brought suit to recover on its claims.

The trial court first examined the provisions relevant to the timing of the contractor's notice. The court found that the contract between the County and the contractor provided that no claim for changed or extra work could be made against the County unless it was notified of the contractor's intention to present such claim "within ten days of the event, thing, or occurrence giving rise to the alleged claim." Similarly, state and local laws required the contractor to provide written notice of its intention to file a claim within ten days of the occurrence or beginning of the work upon which the claim was based.

The court next examined when the "event" or "occurrence", which triggered the contractor's duty to notify the county of its intention to present a claim for extra or changed work, happened. The County argued that the contractor was required to notify the County even before a dispute arose — in effect, simultaneously with its submission of a proposed change order in response to a request or directive in the field. The contractor took an opposite view, arguing that it was entitled to wait until a full-fledged dispute had arisen, and it was able to precisely determine the monetary impact of the extra or change, before having to provide notice of its intention to submit a claim.

The court rejected both arguments, instead concluding that, as to each claim, the triggering event for the notice requirement was the contractor's learning that its proposed change order, or other request for adjustment of contract terms, was denied, disallowed, or disapproved, in whole or in part, by the County. Applying such standard, the court concluded that the contractor did not give the County timely notice of its intent to submit all 93 disputed claims, and thus held that those claims were barred.

Contractors should always be careful to comply with all notice requirements in their contracts. In order to do so, contractors need to have (1) a firm grasp on the time period(s) in which they must take action to preserve their claim and (2) a good understanding of the point of which those periods will begin to run.

About the Author: Katz & Stone is a nationally recognized law firm, with a practice emphasizing construction law and risk management. For the balance of the newsletter from which this article was first published, visit http://www.katzandstone.com/newsletters/dec2003.html. For more information contact: Gerald Katz, Esq., Katz & Stone, 8230 Leesburg Pike, Suite 600; Vienna, VA 22182; (703) 761-3000.

4.4 Absent Contractual Allocation of Delay Risk, Subcontractor Cannot Recover Damages from General Contractor who was not Responsible for their Occurrence (Vol. 5, No. 7)

By: *Gerald Katz, Esq.*

In *In Re: Regional Building Sys., Inc. v. The Plan Comm.*, 320 F.3d 482 (4th Cir. 2003), a federal court held that under New York law, absent a contract term to the contrary, a subcontractor cannot recover delay damages from its general contractor when the general contractor is not responsible for the delay.

In the underlying dispute, a contractor entered into a contract with the owner of a tract of real property in Staten Island, New York to manufacture, deliver, and install modular housing units on the property. Shortly thereafter, the contractor subcontracted a portion of its work. Under the terms of the subcontract, the subcontractor was required to pick-up the contractor-built housing units from the contractor, transport the units to the project site, and then erect and complete the structures.

Several months into construction, however, the owner experienced financial difficulties and defaulted on a number of payments. Without these payments, the contractor experienced severe cash flow problems. As a result, the contractor could not meet its subcontract obligation to deliver the requisite number of housing units to the subcontractor.

The contractor suspended work under the subcontract, forcing the subcontractor to bear the expense of supporting idle labor and equipment. Almost a year and a half into the project, the owner ceased paying the contractor altogether. Finally, after the owner defaulted on two consecutive payments, the contractor suspended deliveries and thereafter terminated both its prime contract with the owner and its subcontract with the subcontractor.

The contractor then filed for bankruptcy, and the subcontractor brought claims before the bankruptcy court, seeking to recover not only payment for the work it had performed but also to recover certain other expenses, additional damages, delay damages, and interest. In the middle of the trial the contractor's bankruptcy trustee agreed to pay the subcontractor $718,128.23, which reflected the undisputed portion of the subcontractor's claim for work actually performed.

As no agreement could be reached as to the remaining claims, however, the parties went to trial. In the end, the bankruptcy court held that the general contractor acted reasonably in suspending the work and ultimately terminating its contracts and that, while the subcontractor could recover a relatively nominal amount for its reimbursable expenses and additional damages, there would be no recovery of delay damages and interest.

The subcontractor appealed the bankruptcy court's holding that it could not recover delay damages from the general contractor, arguing that the contractor's financial difficulty did not excuse performance of the subcontract. The United States Court of Appeals for the Fourth Circuit disagreed, finding that under New York law, absent a contractual commitment to the contrary, a contractor cannot be held responsible for delay damages incurred by its subcontractors unless it caused or controlled the delay at issue. In short, unless there is a specific allocation of the risk of delay, a general contractor does not impliedly guarantee that its subcontractors will not be delayed by factors outside its control.

Turning to the specific facts at issue, the court noted that the contractor's suspension of work was caused by the owner's failure to pay the contractor. As a result, the contractor was not responsible for the delays its subcontractor experienced and was not liable for the subcontractor's delay damages.

Although the court in Regional Building Systems, Inc. applied principles of New York law, its holding reflects the general tendency of courts to leave the cost of a realized commercial risk where it lies in the absence of an express, agreed-upon allocation of that risk.

About the Author: Gerald I. Katz, Esq., KATZ & STONE, L.L.P., 8230 Leesburg Pike, Suite 600; Vienna, VA 22182;phone: 703.761.3000; e-mail: gkatz@katzandstone.com.

4.5 Equitable Adjustment Allowed for Deductive Change Despite Contractor's Unbalanced Bid
(Vol. 5, No. 5)

By: *Philip R. White, Esq.*

Where a public owner issued a deductive change order, it was required to equitably adjust a contract despite the absence of an equitable adjustment clause in the contract, despite the absence of specifications or applicable public contracts law, and despite the fact that the contractor's bid was unbalanced . In *M.J. Paquet, Inc. v. NJ Dept of Transportation*, 171 N.J. 378, 794 A.2d 141 (2002), the NJDOT awarded a contract to Paquet for the rehabilitation of several highways and bridges. A year later, OSHA revised regulations that affected Paquet's bridge painting work. When the NJDOT and Paquet could not agree on a increased price for the painting that resulted from the revision of the OSHA regulations, the NJDOT deleted bridge painting from Paquet's work scope.

Paquet's bid was unbalanced due to a last minute change in painting subcontracts that reduced the cost of bridge painting. Paquet argued that due to time constraints it could not practically adjust all of the pay items affected by the change in subcontractors in time to submit its bid to NJDOT. Instead, Paquet reduced the price of several other items, like mobilization and layout costs, to offset the overstated bridge painting pay items.

After finding that NJDOT correctly deleted the painting work from the contract pursuant to the principle of impracticability, the Court held that Paquet was entitled to an equitable adjustment to compensate it for the non-painting work included in bridge painting pay items that NJDOT had deleted from the contract. The court found that neither the contract nor the specifications contained a provision addressing equitable relief. The court expressly rejected the argument that Paquet's claims was barred by a specification prohibiting claims for additional compensation arising from pay items that inaccurately stated the cost or profit associated with that item.

The court noted that there was an established policy against unbalanced bids and front-end loading in public contracts. But in examining the facts relating to Paquet's bid, the court found Paquet had not violated that policy.

Although the court commented that its ruling was limited to the facts of the case, it changed two basic aspects of public contracting law in New Jersey and provided a basis for contractors to argue for the same changes elsewhere. First, after canvassing federal law and the law of several states that recognize a contractor's right to equitable adjustment by statute or contract, the court determined that New Jersey would permit such relief. Thus, because there was no statutory or contractual basis for equitable relief in the Paquet contract, the court effectively created a common law right to equitable adjustment.

This common law right to equitable adjustment creates a potentially potent tool for contractors to use against public and private owners. Second, rather than simply barring Paquet's claim because of the policy against unbalanced bids, the court undertook an analysis of the reasons why the bid was unbalanced. This gives aggrieved contractors a valuable tool for pressing their claims.

The court's decision seemed driven by its finding that, absent an equitable adjustment, the NJDOT would be unjustly enriched at Paquet's expense. If the court denied relief to Paquet, the contractor would not have been compensated for substantial work that it performed. Viewed expansively, the decision recognized the equitable principle that pricing for changes caused by the owner should be adjusted to make the contractor whole.

About the Author: **Philip White** is an attorney with Sills, Cummis, Radin, Tischman, Epstein & Cross, located at One Riverfront Plaza, Newark, NJ 07102 (973) 643-7000. E-mail: pwhite@sillscummis.com. This article was originally written by Mr. White for publication in the URS *Claims Resource* newsletter, Spring 2003 edition, published by URS Dispute Resolution Group, 100 California Street, Suite 500, San Francisco, CA 94111-4529. For more information on URS, contact Adam Winegard at 213-996-2579 or by e-mail at dispute_resolution@urscorp.com.

4.6 Proving Constructive Acceleration Claims Can be Difficult for Contractors, (Vol. 6, No. 7)

By: *Katz & Stone, L.L.P.*

When claiming constructive acceleration, contractors must prove that any time extensions received were inadequate to remedy their excusable delays. Time extensions do not have to be granted immediately and the mere failure to grant extensions does not, by itself, constitute constructive acceleration so long as the procedures set forth in the parties' contract for time extensions are being followed.

In *Fraser Construction Co. v. United States*, 384 F.3d 1354 (Fed Cir. 2004), the government contracted with a contractor to excavate material from the bottom of a shallow lake. The contract period was from May to September of 1993. To facilitate excavation, the water level of the lake was lowered, leaving only a small stream running through the lakebed which the contractor diverted with a dike. The contractor designed the dike to withstand water flow substantially higher than the average water flow for the lake.

Shortly after the project began, the lake began experiencing high water flows because of rain in the region.

The high water flows overran and destroyed the dike and caused damage to the work site. The contractor experienced delays associated with repairing the dike and inundation of the work site. The contractor requested a time extension which the government denied but later granted. The contractor contended that the extensions that were later granted were not sufficient to compensate it for the extra expenses it had incurred in dealing with the high water flows, because upon being told that it would not receive time extensions or that those extensions would be dealt with later, it was forced to continue its operations at a substantial additional cost. The contractor further argued that many of the time extensions that it was ultimately granted were of no use to it because they were not granted on a timely basis.

The court rejected the contractor's argument that the extensions were not timely, finding that it was standard practice for parties to negotiate after the fact to determine the number of days to extend the contract period. The court found it significant that the government followed the standard procedure for reviewing extension requests and had daily discussions with the contractor regarding delays and progress on the project.

The court also rejected the contractor's claim that the government forced it to accelerate its performance by pressuring it to work through delays that should have resulted in an extension. The contractor's evidence of constructive acceleration consisted of a letter from the government urging the contractor to adhere to the contract schedule and threatening termination if it did not. An expression of concern about progress combined with a refusal to issue extensions can be the equivalent of an order to accelerate. However, this letter was sent by the government before the contractor ever made a claim for excusable delay based on high water flow. Also, the delays which precipitated the government's letter were due to subcontractor problems not associated with high water levels.

As a result of this case, contractors should be aware that when making claims for constructive acceleration, they must prove that extensions granted by the government were not sufficient to offset excusable delays. Furthermore, time extensions do not have to be granted immediately and the mere failure to grant extensions does not, by itself, constitute constructive acceleration so long as the procedures set forth in the parties' contract for time extensions are being followed.

About the Author: This article is reprinted with permission from **Katz & Stone, L.L.P.** Construction Newsletter (March/April 2005 issue). Katz & Stone's practice is devoted to the construction industry, with its attorneys working exclusively for those who own, develop, design, build and bond construction projects. The address is 8230 Leesburg Pike, Suite 600, Vienna, VA 22182. (703) 761-3000.

4.7 When to Stop Work for Non-Payment (Vol. 5, No. 3)

By: Susan Linden McGreevy, Esq.

Every subcontractor has had to deal with slow payment by general contractors who claim that it's not their fault – the customer hasn't paid them and, in a slow economy, this refrain is heard more and more frequently.

As sympathetic as the sub might feel, the unfairness of being held up due to a problem you had nothing to do with is more than just irritating, especially when you've already had to pay the bills for that work. The temptation to just stop working is hard to resist, even though general contractors may be making serious threats and pointing to contract clauses that (they say) require the sub to stay on the job.

A Virginia drywall contractor's frustration caused him to take the chance, and it ended him up in court. Manganaro Corp. was a sub to HITT Contracting for a Lucent Technologies project in downtown Washington . Although Manganaro's subcontract had a strongly worded "pay if paid" clause ("Contractor's receipt of payment by the owner shall be a condition precedent to the obligation of the Contractor to make any payment to the Subcontractor"), Manganaro had had the foresight to negotiate in another sentence that took most of the teeth out of this clause: "Notwithstanding the above, it is understood the Contractor has the ultimate obligation to pay the Subcontractor within a reasonable time regardless of payment status from the Owner."

When payment was not forthcoming – due to non-payment by Lucent – after much discussion and correspondence, Manganaro suspended work. HITT got another contractor in to complete the work and back-charged Manganaro. The next step was federal court.

The judge found first that while a "pay if paid" clause might be enforceable in Virginia , this clause was not, given the extra language Manganaro had negotiated into the contract.

The next question was: "What's a reasonable time for payment?" HITT argued that Manganaro had not given it enough time to at least try to collect from Lucent, but the court decided that Manganaro didn't have to wait at all. Accepting Manganaro's testimony that 30 days from invoice was a "reasonable" amount of time, the court ruled that HITT's persistent failure to pay within 30 days was unreasonable. Although there were no Virginia court decisions directly on this point, "the authorities uniformly state that a subcontractor who is unreasonably denied payment as he progresses towards completion is justified in suspending performance until he is paid." *Manganaro Corp.*

v. *HITT Contracting Inc.*, 193 F. Supp.2d 88, 96 (D.D.C. 2002).

HITT argued that even if it was in breach of contract (and it tried hard to convince the court that its delay in payment hadn't been all that bad), not every breach of contract justifies the other party's refusal to perform. The court agreed that this had been the law in Virginia for more than 100 years, and that some breaches can be remedied by a payment of money damages after the fact – such as interest, for late payments.

Ironically, one of the reasons why the court went against HITT on this point was that HITT had inserted in its contract the sentence, "Time is of the essence." While HITT (and most other contractors) intended this clause to apply to Manganaro's obligation to perform, the court read it as applying both ways: It meant that HITT had to perform its obligation to pay in an expeditious manner.

The case was particularly difficult for the judge to decide, since Manganaro's billings were not entirely accurate, and HITT could (and did) argue that it thought it was making the appropriate payments, or at least partial payments. The court ruled that it had to look at the objective facts of what Manganaro was really due, since HITT's own records would have shown that Manganaro was due $140,079 when it suspended performance.

"HITT's breach of its most fundamental contractual obligation to pay for the work it approved relieved Manganaro of any further obligation under the contract, including the obligation to [do extra work] and to do the repair work on the punch list."

Not every slow-pay situation will justify suspension of work, given that business relations will inevitably be scarred if not totally ruined in the process, and even with interest of $27,753.60 thrown in, Manganaro was probably not made whole for all the expense it incurred in taking this dispute to federal court. Nevertheless, I expect to see the words of this decision quoted by many lawyers for subcontractors in letters and legal briefs on the subject of slow payment. And the words just might find their way into some court decisions in other states.

About the Author: *Susan McGreevy* is a partner at Husch & Eppenberger, Kansas City, Mo., 816/421-4800, e-mail to susan.mcgreevy@ husch.com. This article was first published in Contractor Magazine, the Newsmagazine of Mechanical Contracting and is reprinted with permission. Penton Media, Inc.; 2700 South River Road, Suite 109; Des Plaines, IL 60018; Tel: (847) 299-3101.

4.8 Differing Site Conditions, Defective Specifications: One Coin, Two Sides (Vol. 5, No. 4)

By: *Stephen J. Densmore, Esq*

Unforeseen site conditions typically spawn two types of claims based on two distinct but related theories: differing site conditions (DSC) and defective specifications. A contractor may attempt to circumvent the limitations on recovery under a DSC provision by characterizing its claim as one for breach of contract due to defective specifications. In *Comptrol Inc v. United States* (Fed Cir. 2002) 294 F.3d 1357, a federal court recognized the close relationship between these two theories and blocked the contractor's attempt to do an end-run around the recovery limits in the owner's DSC provision.

In *Comptrol*, the contractor claimed that during bidding, the government withheld material information concerning the presence of quicksand and the location of a subsurface pipeline. Apparently unhappy with the recovery allowed by the DSC provision, the contractor tried to argue that its claim should be permitted as a claim for damages for breach of contract arising from defective specifications. The court rejected this argument and held that although DSC claims and defective specifications claims are distinct in theory, "where the alleged defect in the specification is the failure to disclose the alleged differing site condition," the two claims are "so intertwined as to constitute a single claim" that is governed by the DSC provision."

Comment: DSC provisions typically contain a number of clauses relating to such things as notice, cost recovery, and continuous work that are valuable protections for the owner. In order to preserve these protections, an owner should try to draft the DSC provision with language that limits the contractor's ability to circumvent the DSC provision by characterizing its claim as one for "defective specifications."

About the Author: Steve Densmore is an attorney with the law firm of Heyman & Densmore, 5820 Danoga Ave., Suite 250; Woodland Hills, CA 91367; Tel: (818) 703-9494. Email: steven@hdlawllp.com; Phone: (818)703-9494.

Chapter 5

Contract Language Issues & Concerns

5.1 Boilerplate Can Burn! (Vol. 5, No. 4)

By: *Lawrence Moss, Esq.*

There are no "industry standard" construction form documents. But the forms issued by the American Institute of Architects (AIA) come close, being widely used. One key provision to be considered is the issue of consequential damages.

"Consequential damages" are damages, losses or injuries which do not flow directly and immediately from the wrongful act of a party, but rather from some consequence of that act, such as lost profits or damage to reputation.

Some years ago, in the case of *Perini Corp. v. Great Bay Hotel,* a New Jersey appellate court upheld a fourteen million dollar consequential damages award against a general contractor for the owner's lost profits resulting from late completion – where the general contractor's total compensation was only $600,000!

In response to this and similar cases across the country, the AIA, at the urging of contractors, included the mutual waiver of consequential damage in the 1997 forms (see Paragraph 4.3.10) of the General Conditions to the Contract for Construction, AIA Document A201 (1997); there are analogous provisions in the current Owner/Architect agreement forms.). By eliminating consequential damages, the AIA intended that contractors and architects be protected against damages awards which are grossly disproportionate to the compensation received for their services.

Although the waiver is mutual, owners, contractors and architects need to consider the potential impact of this provision.

For example, consider the development of a luxury condominium project. The owner/developer probably won't see a dime of net income until the project is completed and it starts unit closings. During construction, however, the owner is incurring costs: interest and financing fees; taxes, insurance and other carrying costs; administrative costs, overhead and salary of sales and development personnel. The owner's hope of making money is premised, in part, upon construction being completed in time to get units closes and costs paid or eliminated before those costs eat away any chance of profit – or worse.

If the contractor fails to complete by the promised completion date, the owner will pay interest and carrying costs longer than planned. Equity investors, who have a preferential return on capital, will earn more money, leaving less bottom line for the developer. Buyers might be able to terminate their contracts because of a missed delivery date. Contractors can be delayed, too. Drawings may be incomplete and may need to be revised, resulting in extra delay and expense.

To the extent that these damages are considered "consequential" damages, and a party signed an unmodified AIA contract, that party won't recover a penny.

An alternative to deleting the waiver is to reconsider liquidated damages. Liquidated damages are amounts recoverable in a fixed amount upon occurrence of a defined event. If liquidated damages clauses are drafted properly so as not to be punitive, than after appropriate economic analysis, they are usually enforceable by the courts.

Liquidated damages have been generally anathema to contractors. Since the adoption by the AIA of the mutual waiver language, however, more contractors are willing to negotiate a serious and meaningful liquidated damage remedy -- especially if the alternative is the deletion of the waiver of consequential damages clause, and the possibility of uncapped liability for the contractor.

The moral of the story: If you are presented with an unmodified AIA contract for signature, think twice, because the consequences of signing an unmodified AIA contract could be significant.

About the Author: Lawrence Moss is an attorney with the law firm of D'Ancona & Pflaum LLC, 111 E. Wacker Drive, Suite 2800, Chicago, IL 60601; Tel: (312) 602-2000. The firm's web site is: http://www.dancona.com.

Chapter 6

Damages

6.1 Contractor May be Sued for Lost Profits arising out of Breach of Contract (Vol. 6, No. 5)

When an oral surgeon hired a contractor to construct his office and was unable to use part of the finished office due to problems with the floors, he sued the contractor for damages, including loss of profits. The court held that the possibility that there would be lost profits if the facility were not open and available for business in time could have been reasonably foreseen at the time the parties entered into the contract. The surgeon was entitled to have a jury determine whether lost profits should be awarded as part of the compensatory damages.

After opening his dental practice, the floors in the doctor's office began seeping moisture, becoming slippery, and producing offensive odors. He closed the surgical rooms of his office due to these conditions which were caused by improper ventilation of the concrete slab under the flooring. He sued for lost profits due to lost patients and lost business growth opportunities. The contractor argued that such losses were not included in the measure of damages for breach of a construction contract.

The case law of the state having jurisdiction over this case (Connecticut) supports awarding lost profits as an element of compensatory damages for general breach of contract claims, says the court. Citing the *Restatement (Second) of Contracts*, the court states that recovery is divided into the components of direct damages and incidental or consequential loss caused by the breach, and goes on to state that traditionally, consequential damages include "any loss that may fairly and reasonably be considered [as] arising naturally, i.e., according to the usual course of things, from such breach of contract itself." The court quotes previous case law for the proposition that "it is our rule that unless they are too speculative and remote, prospective profits are allowable as an element of damage whenever their loss arises directly from and as a natural consequence of the breach." *Amrogio v. Beaver Road Associates*, 267 Conn. 148, 836 A.2d 1183 (2003).

Comment: This case demonstrates why contractors and design professionals are seeking waivers of consequential damages in their contracts with project owners. Some of the American Institute of Architects (AIA) standard form agreements contain such waivers to protect the contractor against consequential economic damages such as lost profits or lost rents. Project owners that believe such economic damages may be a significant part of their project risks may find it to strike such waivers of consequential damages out of the contracts for that reason. I have attended more than one construction lawyers program where attorneys for project owners have stated that they routinely strike these clauses from contracts.

Chapter 7

Design-Build

7.1 Book Review: *Design-Build Lessons Learned*
7.2 Ambiguity in Design-Build Contract Specs Interpreted in Favor of Contractor
7.3 Design-Builder Not Entitled to Equitable Adjustment to Meet Owner's Detailed Design Specifications
7.4 Liquidated Damages Clause and Waiver of Consequential Damages Clause Effectively Cap Damages Available Against Design-Builder
7.5 Public Agency Exempted Project from Competitive Bidding

7.1 Book Review: Design-Build Lessons Learned
(2004 Edition), 240 pages (Vol. 7, No. 8)

Those of you who are interested in design-build have undoubtedly come across *Design-Build Lessons Learned*, an annual publication authored by Mike Loulakis, the president of the law firm of Wickwire Gavin and one of the country's foremost authorities on design-build. Mike first started publishing *Design-Build Lessons Learned* in 1995 as a newsletter designed to give his clients an opportunity to understand the nuances of this new area of law. The first edition of the publication only covered ten cases, which represented the universe of 1995 court decisions addressing design-build disputes. Ten years later, after unprecedented growth in both public and private sector use of the design-build process, Mike has now reported on a total of 304 cases, almost half of which have come during the 2002-2004 time period. *Design-Build Lessons Learned* has also become much more substantial in appearance, as it moved from newsletter format to paperback book format in 2001.

As an avid reader of this publication, I strongly believe that the industry needs caselaw to help explain design-build legal theories and precedent. Over the past few years, *Design-Build Lessons Learned* has regularly highlighted 20-30 important design-build cases around the country. While this gives design-build practitioners insight into the mistakes of others, it also affords them the opportunity to fashion best practices from these cases.

The 2004 edition of *Design-Build Lessons Learned* is a 240-page paperback book that contains an array of significant design-build cases among its 58 reported decisions. Although it is very difficult to identify the "best of the best," special attention should be paid to the following seven cases:

- *Lockheed Martin Idaho Technologies Co. v. EG&G Idaho Inc.*, where an Idaho federal court rejected a turnkey contractor's excuses for failing to meet the contract's performance guarantees. Particularly noteworthy was the court's conclusion that millions of dollars of cost overruns were not sufficient to establish a "commercial impracticability" defense to performance.

- *Record Steel and Construction v. United States,* where the United States Court of Federal Claims concluded that the design-builder was entitled to a change order when the government required it to use a "recommended" geotechnical approach instead of what the design-builder wanted to use.

- *PDC-El Paso Meriden, LLC v. Alstom Power, Inc.*, where a Massachusetts state court concluded that an owner and an EPC contractor needed much more than a Letter of Intent to create a binding commitment to construct a power plant.

- *Siemens Westinghouse Power Corp. v. Dick Corp.*, where a New York federal court refused to find that one member of a design-build consortium could claim that it was "duped" by its partner.

- *Metropolitan Steel Industries, Inc. v. Perini Corp.*, where a New York state court considered a series of claims arising from construction of a New York City Transit Authority bus terminal. Particularly noteworthy about this case is the guidance it provides relative to the potential liability of a subconsultant engineer to the prime design-builder, reaching a different result than what we have seen from other cases around the country.

- *O'Brien & Gere Technical Services, Inc. v. Fru-Con Construction Corp./Fluor Daniel, Inc.*, where the 8th Circuit Court of Appeals concluded that a design-build contract was abandoned because of, among other things, poor change order management and a failure of the parties to ever reach agreement on a design baseline.

- *American Family Mutual Insurance Company v. American Girl, Inc.*, where the Wisconsin Supreme Court finally ended a multi-year piece of litigation by concluding that a design-builder's CGL insurance policy covered the damages incurred when a building failed because of negligence by the design-builder's geotechnical contractor.

For an extensive excerpt of one of the cases taken directly from the publication, see the next article in this issue of the ConstructionRisk.com Report.

7.2 Ambiguity in Design-Build Contract Specs Interpreted in Favor of Contractor (Vol. 7, No. 8)

By: *Michael C. Loulakis*

Record Steel and Construction v. United States, 62 Fed. Cl. 508 (2004), provides an excellent example of the evolution of design-build caselaw. The dispute in this case involved whether a design-build contract required foundations to be over-excavated. The design-builder argued that the contract unambiguously made over-excavation a design recommendation – not a design requirement. In the alternative, it argued that if the contract was ambiguous, then the ambiguity was latent and should be construed against the government. The government argued that the contract expressly and unambiguously required the design-builder to over-excavate the foundation. After carefully examining the relevant contract provisions, the United States Court of Federal Claims ("Court of Federal Claims") found the contract to be latently ambiguous, and saddled the government with the financial responsibility of the over-excavation. [Editor's Note: This casenote discussion is excerpted from *Design-Build Lessons Learned (2004)* - See book review above for more details].

The Corps of Engineers ("Corps") awarded Record Steel and Construction, Inc. ("Record Steel") an $8.8 million design-build contract for a dormitory at Offutt Air Force Base in Bellevue, Nebraska. Part of the RFP contained a foundation analysis report, with a section entitled

"Subsurface Recommendations." Included in the recommendations was the following language:

> Due to the anticipated column loads for a multi-story building, it is believed that improving the site is more viable than reducing the bearing pressure to a very low value .... The recommended improvement program is outlined below.

The recommended program contained statements that materials be undercut and "should be excavated" from below the bottom elevation of all building footings.

In response to the RFP, Record Steel submitted a price proposal informing the Corps that it did not believe over-excavation for the foundations would be required but, if site conditions ultimately required over-excavation, Record Steel committed to perform this work at no additional cost. The need for over-excavation was discussed during several design meetings both prior to and after contract award. The parties agreed that Record Steel's geotechnical firm was to conduct field investigations and tests and provide such information to both Record Steel and the Corps. If the resulting data was satisfactory, then Record Steel could proceed with its design without conducting over-excavation.

The geotechnical firm concluded that the native soils were adequate to support the building's footings without over-excavation. However, the Corps apparently re-evaluated its position and refused to issue a notice to proceed for the footings unless Record Steel agreed to conform to "the requirements of the subsurface recommendations of the Foundation Analysis Report" and over-excavate the site. Record Steel complied with this order and submitted a claim for approximately $188,000 for the costs associated with the over-excavation effort.

The Court first looked at the reasonableness of each party's contract interpretations. In finding that Record Steel's interpretation was reasonable, it first noted that Record Steel, as the designer-of-record, was expected to exercise its professional judgment in designing the dormitory and had to defer only to specific requirements contained in the RFP, not to recommendations. The Court then examined how the "requirements" in the RFP were expressed in terms of words like "shall," "may," and "should."

It found that the most critical aspects of the foundation report used the word "should" instead of "shall" – and that this expressed a desire for action, but not a binding requirement. It looked to the fact that the foundation report stated that the Corps "believed" that over-excavation was "more viable" to improve the site, and couched its report in terms of a "recommendation" rather than as a requirement. The Court also found Record Steel's interpretation to be reasonable based on the fact that the Corps' initial borings were not conducted within the actual footprint of the dormitory's location.

The Court ruled, however, that the Corps' contract interpretation fell "within the zone of reasonableness." It looked to the fact that the RFP used the verb "shall" in connection with incorporating the foundation report's recommendations into the contract, and that, by referring to the terms "over-excavation and compaction requirements," there was an argument that the RFP expressly converted the foundation report's recommendations into requirements.

Faced with two reasonable contract interpretations, the Court then looked to the rule of *contra proferentem* for guidance on who should bear the risk of these ambiguities. The four-part test associated with this rule places the risk of the ambiguities on the government when: (1) the contract specifications were drawn by the government; (2) the language used therein was susceptible of more than one interpretation; (3) the intention of the parties does not otherwise appear; and (4) the contractor actually and reasonably construed the specifications in accordance with one of the meanings of which the language was susceptible. The Court found that all of these conditions were satisfied.

The court also refused to apply the exception to the general rule of *contra proferentem* (i.e., the patent ambiguity doctrine), which resolves ambiguities against the contractor where the ambiguities are "'so 'patent and glaring' that it is unreasonable for a contractor not to discover and inquire about them.'" The court did not find this ambiguity obvious, particularly since the Corps had not indicated its view of the mandatory nature of these so-called "requirements" until many predesign meetings between the parties had taken place.

There are several "take-aways" from the *Record Steel* case. Owners who use competitive procurement processes, like the Corps of Engineers and other public agencies, must remember that proposers get the benefit of the doubt if there is something wrong with or unclear about the owner-furnished RFP documents. The fact that the over-excavation concept was expressed in terms of a "recommended" process started the potential for ambiguity, and the "should," "may," and "shall" verbs contained in the foundation report sealed the deal for the conclusion that there was an ambiguity. If the Corps wanted to mandate over-excavation, it should have said so explicitly.

Equally important, however, is that Record Steel did the right thing by advising the Corps during the proposal phase that it did not intend to undercut unless soil conditions required it to do so. These actions demonstrated that Record Steel actually relied upon its interpretation of the ambiguity, which is one of the predicates to recovery under the *contra proferentem* rule. The result of this case might have been different if Record Steel recognized the ambiguity but never informed the Corps of its interpretation during the proposal process.

7.3 Design-Builder Not Entitled to Equitable Adjustment to Meet Owner's Detailed Design Specifications

(IRMI Expert Commentary, K. Holland - September 2004)

Where a contractor/bidder has actual knowledge of an ambiguity in the specifications included within the request for proposals on a design-build project, the bidder has an obligation to inquire about the ambiguity. This is true whether that ambiguity is so obvious that it is "patent" or it is so unapparent at time of bidding that it is "latent." Where a design/builder failed to inquire about the discrepancy, a court held it was not entitled to recover an equitable adjustment claim because it unilaterally chose how it would resolve the ambiguity without first inquiring of the project owner.

The case was *United Excel Corp.* (VABCA #6937, 2003 WL 22977508, 041BCAP32, 485). There, a design-build contractor, United Excel Corporation (UEC), submitted its 90 percent design submission to the project owner, Veteran's Administration (VA). The VA representatives formally commented that the request for proposal (RFP) required stainless steel operating room heating, ventilation, and air-conditioning (HVAC) components. UEC stated at that time that it would revise the drawings to reflect the stainless steel instead of the aluminum components it had included in the design at that point. In all subsequent shop drawing submittals, the stainless steel components were included and were approved by the VA.

UEC submitted two change order proposals on behalf of its subcontractors. As explained by the court, the VA denied the change orders because:

> UEC submitted a certified claim for an equitable adjustment resulting from installing stainless steel instead of aluminum diffusers. The VA conceded that the specifications contained conflicting provisions concerning whether stainless steel or aluminum was to be used for the diffusers in the operating room HVAC installation. But the VA argues that the discrepancy was "patent" and as such that the D/B was required to inquire about what material was required.

The request for proposal (RFP) stated: "The RFP documents are intended to define existing conditions, certain required items, and design parameters to be included in the project. It is the DB Team's responsibility to complete the documents and construction in a manner consistent with the intent of the RFP documents within the required time period (contract length)."

In discussing this issue, the court pointed out that if an ambiguity is indeed so obvious or glaring as to be a "patent ambiguity," then the rule of *contra proferentum* (interpretation against the drafter) would not apply. But this did not help the design-builder in this instance because its subcontractors (Stadium/Triangle) were aware of the specification discrepancies with regard to the material requirements prior to submitting their proposal to the design-builder.

As explained by the court, "Where a contractor/bidder has actual knowledge of an ambiguity, be it 'patent' or 'latent,' it has an obligation to inquire about the ambiguity." Since the design-builder failed to inquire about the discrepancy, the court held it could not now prevail on an equitable adjustment claim resulting from its unilateral resolution of the ambiguity.

In addition to the above argument, the design-builder argued that because it was working under a design-build contract, the drawings and specifications contained with the RFP must only be "design parameters" and not absolute requirements. The D/B, therefore, asserted it was entitled to choose aluminum diffusers as the most economic way to achieve the design intent. In rejecting this argument the court explained:

> The contract is clear that, in executing the final Construction documents, UEC was constrained to follow the requirements of the RFP specifications and drawings, and this constraint required UEC/HWA to design a diffuser configuration, using stainless steel diffusers ... We also see nothing in the case law ... for the proposition that the well-settled law relating to the contract interpretation is suspended or abrogated in a design-build contract. To the contrary, the case law indicates that a design build contract shifts risk to a contractor that a final design will be more costly than the bid price to build and that the traditional rules of fixed-price contract interpretation still obtain. UEC was not relieved of its obligation to inquire about the aluminum stainless steel diffuser discrepancy because the Contract was design-build.

One final argument by the design-builder that was rejected by the court was the assertion that the court should create a new method of contract interpretation for design-build contracts because use of the traditional "patent ambiguity" rules of interpretation "unduly punish" contractors where a contractor bids on incomplete plans and specifications as is typical with design-build projects. The court concluded:

... there is nothing in the terms of the Contract or the law that would permit us to ignore the Contract language and establish a new rule of allocating the risk that a patent ambiguity exists in the specifications of a design-build RFP.

Comment: The necessity of a design-build contractor meeting design criteria specified by the project owner has been addressed by a number of court decisions. In a number of cases, contractors have argued that the owner's design details were not binding on them because they believed a design-build contract permitted them to disregard such details so long as what they designed met the performance requirements of the owner. But courts hold that owners have the right to prescribe plans and specifications just as detailed as in design-bid-build projects. When responding to a design-build solicitation, the contractor needs to understand which aspects of concept documents developed by the owner are discretionary and which are not.

A good example of this is presented by the case of *Dillingham Construction v U.S.*, 33 Fed. Cl. 495 (1995). The facts underlying that case were that the electrical specifications included in the solicitation required use of raceways (trays) to run conduit and described conduit size and characteristics as well as supports for the raceways. Dillingham's electrical subcontractor wanted to use metal clad cable instead of the raceways. The owner rejected this proposal. Then, when the subcontractor installed the raceways using supports differing from those specified, the owner required them to be removed and replaced.

The subcontractor submitted a claim for over $600,000 for its extra costs in complying with these requirements. It argued that the specifications were "performance" specifications providing "general guidelines" giving the subcontractor "wide latitude" in interpreting them. The court rejected the subcontractor's argument and noted that the contract specifically required the subcontractor to furnish a design that complied with the electrical specifications. The court found the specifications to be "design" specifications that gave the subcontractor no flexibility to deviate.

Just as the contractor is required to meet the significant design details that are specified by the owner, the more involved the project owner becomes in specifying such details, the more the owner takes responsibility for the design, the more involved it becomes in specifying such details.

7.4 Liquidated Damages Clause and Waiver of Consequential Damages Clause Effectively Cap Damages Available Against Design-Builder

(IRMI Expert Commentary, K. Holland - March 2003)

The design-build case, *Mistry Prabhuda Manji Eng. Pvt. Ltd. v Raytheon Engineers & Constructors, Inc.*, provides insight into the judicial interpretation of contract clauses that purport to limit liability of engineers/contractors. In commercial settings, explained the court, a limitation of damage clause will rarely be found unconscionable in the absence of oppression and unfair surprise. Risk managers need to take note that, generally speaking, courts will enforce the terms of the contract that result from arms' length negotiations between two commercial entities.

This article examines a recent design-build case, *Mistry Prabhuda Manji Eng. Pvt. Ltd. v Raytheon Engineers & Constructors, Inc.*, 213 F. Supp. 2d 20 (US DC, Mass 2002).

The Facts: Contracts requiring a design-build engineering firm to supply "basic engineering packages" for licensing and technology transfer agreements for the design and construction of a processing plant for sodium hydroxide (caustic soda) contained a liquidated damages clause capping the engineer's liability at 10 percent of its fee. They also contained a waiver of consequential damages clause waiving "special, indirect, incidental, or consequential damages of any kind." In response to the project owner's suit against the engineer for failure of the plant to achieve commercial production, the court enforced these clauses to limit the available recovery.

The plaintiff's complaint against the contractor alleged breach of contract, misrepresentation, and fraud. With regard to the counts of the complaint alleging misrepresentation and fraud, the court dismissed these because they were barred by the 2-year statute of limitations. In response to the defendant's argument that the breach of contract claim should also be dismissed based upon the waiver of consequential damages and the liquidated damages clauses, the plaintiff argued that the clauses should not be enforced because the clauses were unconscionable, were based on material misrepresentations, and were the product of mutual mistake.

The waiver clause provided: "Article XV Waiver of Consequential Damages. In no event shall Seller [contractor] be liable to [owner] whether in contract, warranty, tort (including negligence or strict liability) or otherwise for any special, indirect, incidental or consequential damages of any kind or nature whatsoever."

The liquidated damages clause provided:

> Article VIII Liquidated Damages. In the event that the Caustic Prill Unit fails to produce Caustic Soda beads during the performance test even though all the conditions described in Article VII hereof have been satisfied and despite [contractor's] efforts to correct said failure, for each 5 percent or part thereof shortfall below the

level warranted in Article VII, hereof, [contractor] will pay to [owner] an amount equal to 5 percent of the lump sum fee received by [contractor] for the failed Caustic Prill Unit. However, [contractor's] maximum limit of liability under the Agreement as to any failed Caustic Prill Unit shall be 10 percent of the Lump sum fee received by [contractor] for the failed Caustic Prill Unit. These payments are the exclusive remedies provided to [owner] under this Agreement. Except as provided in the Article VII, Contractor shall have no other liability whether in contract, warranty, tort, or otherwise.

The plaintiff, project owner, tried to get around the liquidated damages clause by arguing that it only applied in the event that the Unit failed the performance test. Since there was never a performance test, it argued the limitation clause had no effect.

The Ruling: In interpreting the contract on this matter, the court explained that "the intention of the parties is a paramount consideration." Intent must be ascertained from the contract document itself when the terms are clear and unambiguous. The court concluded that the clause makes clear that although the 5 percent cap appears to apply in the event of a performance test failure, the 10 percent cap applies to any claim under the Agreement regardless of whether or not performance tests were performed. The court emphasized that "When combined with the extremely strong liability-limiting language of the entire clause, these phrases make clear that the intention of the parties was to limit [owner's] recovery under any circumstance to ten percent of the fee it paid to [contractor]."

The court also rejected the project owner's argument that the clauses were "unconscionable" and should not be enforced. The court said that the test under Pennsylvania jurisprudence for unconscionability is "an absence of meaningful choice on the part of one of the parties together with contract terms which are unreasonably favorable to the other party." It further explained that the principle underlying the concept is to prevent oppression and unfair surprise, but that it is not intended to disturb the "allocation of risks because of superior bargaining power."

In other words, just because a party has greater bargaining power and negotiates a more favorable and even onerous deal does not make the deal unconscionable in the absence of oppression and unfair surprise. In commercial settings, explains the court, a limitation of damages clause will rarely be found unconscionable.

In this case, the owner claimed that it was a small unsophisticated Indian company that trusted "an American behemoth" when its president flew to Philadelphia to sign the deal. It made no changes to the contract and did not seek counsel to assist with its negotiation. Although the court described this as a "sympathetic picture," the court concluded that the scenario did not suggest any lack of meaningful choice. In its conclusion with regard to this issue, the court said

> There is nothing in the record to suggest unfair surprise.... The clauses were not hidden boilerplate. The one point which gives this court pause is whether a 10 percent cap creates an adequate incentive to perform. However, there is no indication that the profit margin was any higher than 10 percent. Therefore, [owner] has not demonstrated unconscionability.

Comment: This case provides valuable insight into the judicial interpretation and application of contract clauses that purport to limit liability of engineers and contractors. There is a striking similarity in the project owner's arguments with those that have been raised in so many other reported cases. This decision should be a reminder to every commercial entity entering a contract for the design or construction of a project that, generally speaking, courts will enforce the terms of the contract that result from arms' length negotiations between two commercial entities. This is true even if one of the parties was significantly smaller than the other and did not have equal bargaining clout.

The key, as explained by this court, is whether the damage limitations would be unconscionable. In my own legal practice, I have had more than one client tell me that they wanted to ignore my advice and sign onerous contracts in which they would to be giving up substantial rights with the expectation that they could convince a court that they signed the contract as a result of duress or unequal bargaining position, and that the clause should be found void as unconscionable or contrary to public policy.

My advice has been that a court would not be impressed with their arguments for much the same reasons stated by the court in this case. Plus, my clients have had competent legal assistance with their contracts and this makes their chances of getting a court to let them out of a bad deal even more unlikely. Note, however, that the court provides significant pointers in drafting an enforceable limitation of liability clause, when it states that the clause in this case was not "hidden boilerplate" and that the question of whether a 10 percent cap creates an adequate incentive to perform gave the court pause.

I typically advise clients to make clauses such as indemnification, limitation of liability (LoL), and waiver of consequential damages clear and pronounced in the contract. If an LoL clause might be subjected to close judicial scrutiny, it may even be advisable to have your client separately initial or sign their name beside the clause so they cannot later claim they were surprised to learn of its presence in the contract.

In addition, you should be careful to make the LoL amount reasonable. If it is too small in comparison to the size of the fee or the significance of the potential damages that could occur, a court may refuse to enforce it. Most important of all, the decision of this court demonstrates the value of seeking contract language to limit the liability or the types of damages that can be recovered.

7.5 Public Agency Exempted Project from Competitive Bidding

(IRMI Expert Commentary, K. Holland - February 2002)

In the case discussed in this article we will review how a recent court decision may offer useful instruction for design-builders interested in proposing innovative projects that might not be possible under traditional public procurement methods.

When the Tri-County Metropolitan Transportation District in Portland ("Tri-Met") decided to exempt from competition the contract to construct a light-rail extension to the airport, the Associated Builders and Contractors (ABC) filed suit, challenging Tri-Met's decision. The trial court granted summary judgment in favor of Tri-Met, finding that Tri-Met's decision to forego competition was supported by findings that were based on substantial evidence satisfying the statutory requirements for waiving the competitive bidding requirements.

This decision was affirmed on appeal, with the appellate court providing a detailed explanation of the statutory scheme and basis for upholding the Tri-Met decision. The decision may offer useful instruction for other jurisdictions and design-builders interested in proposing innovative projects that might not otherwise be possible under traditional public procurement methods.

Oregon Law

Under Oregon law, public contracts must ordinarily be awarded on the basis of competitive bids. Section 279.015 of the Oregon Statute authorizes public agencies to exempt contracts from competitive bidding if local contract review boards make findings that certain conditions have been met. In this case, Bechtel Infrastructure Corporation proposed that Tri-Met and two other agencies form a partnership to construct a rail extension to the airport. Bechtel agreed to take responsibility for the design and construction and also to provide project funding in exchange for a fee plus real estate development rights along one segment of the line. Tri-Met evaluated the proposed contract and determined that it met the statutory standard for an exemption. The findings were submitted to a contract review board ("Board") for review and approval, and the Board approved the decision.

In arguing against the exemption, the ABC submitted evidence that Bechtel intended to perform the contract under a national agreement with unions, with the result that nonunion subcontractors would be effectively prevented from competing for work on the project. In response, Tri-Met stated that it was not its policy to dictate terms of the labor relationship between a general contractor and its subcontractors. In addition, it set forth findings that the project could only be built via the means proposed by Bechtel.

Under the statute, a contract could only be exempted from competitive bidding requirements if the public agency finds:

1. It is unlikely that such exemption will encourage favoritism in the awarding of public contracts or substantially diminish competition for public contracts; and
2. The awarding of public contracts pursuant to the exemption will result in substantial cost savings to the public contracting agency. In making such finding, the director or board may consider the type, cost, amount of the contract, number of persons available to bid and such other factors as may be deemed appropriate.

The Appellate Court's Ruling

On appeal, the ABC offered three arguments against the Board's decision to approve the exemption. It argued that even if an agency may exempt a contract from competitive procurement, it must nevertheless assure that some other form of competitive procurement occurs. Second, it argued that before exempting a contract, the agency must consider the effect it would have on subcontractors that want to work on the project. Finally, it argued that there were insufficient findings of fact to support the decision. In rejecting each of these arguments, the court's holding clarifies the Oregon law and could provide guidance for design-build projects in other jurisdictions as well.

With regard to using some alternative method of assuring competition, the court held that the statute did not require this. It only required the use of alternative contracting methods "when appropriate" in the opinion of the Board. Since the Board in this case determined that the only way to award this contract was by sole source to Bechtel because "the Board findings make clear that no one else could have done the job," the court was satisfied that the Board did not err by declining to direct an alternative contracting method that would have included some form of competition.

Concerning the ABC's argument that the Board should have considered how Bechtel's participation in the national labor agreement would effectively preclude nonunion subcontractors from competing for work on the project and reduce competition among subcontractors, the court held that the subcontracts were not considered "public contracts" under the statute, and that "the Board did not need to consider the effect that awarding the contract to Bechtel would have on competition among subcontractors...."

In rejecting the ABC's argument that the Board's findings were not presented in sufficient detail and form, the court held that no particular form of decision is necessary, and that the findings themselves were more than adequate to support the decision.

The Contract Review Board Findings

Specific Board findings included the following:

1. The terms and conditions of the design/build contract will be the result of arm's-length contract negotiations between Tri-Met and Bechtel, and will be reviewed and approved by the Tri-Met Board, thus discouraging any favoritism in the awarding of the contract.
2. Under the unique circumstances presented, it is unlikely that an exemption authorizing Tri-Met to negotiate a design/build contract with Bechtel for the Airport Max Project will encourage favoritism in the awarding of public contracts or substantially diminish competition for public contracts.
3. This negotiated contract will have no negative impact on the contracting or subcontracting market in the Portland area. A sufficient number of qualified firms are available to ensure adequate competition and any subcontracted work will be competitively bid. This is a unique opportunity to enhance the local economy because it is unlikely the Project would be built in the near future without this public-private cooperation.
4. Although the contract with Bechtel will be awarded without direct competition, a substantial part of the construction work will be performed by subcontractors who will be selected through a competitive process. Because of the magnitude of the private investment required and private economic risks associated with the project, the project is unique and it is highly unlikely that this procurement will have any effect or substantially diminish competition for future public contracts.

Based on its review of the record, the court concluded that substantial evidence supported the Board's findings. The decisions of the Board and trial court were, therefore, affirmed.

Conclusion

This decision should be encouraging to public agencies that may be looking at the potential for using public-private partnerships to obtain necessary private financing for projects directly benefiting the public. Design-build contracting scenarios that include private investment, and even ownership interests by the design-builder may offer public agencies the opportunity to construct projects that, as in this case, otherwise would be unlikely to "be built in the near future without this public-private cooperation."

In the current economy, such project financing scenarios that include the design-builder's participation may become increasingly significant. It is important, therefore, that the decision in this case recognized this as a legitimate consideration of Tri-Met.

Another aspect of this decision that might have significance beyond the specific facts of this case is the conclusion that "the fact that the public agency supplies funds to general contractors, who use them to pay subcontractors, is not sufficient to convert subcontracts between private parties into public contracts." This seems to have the potential to give greater leeway and flexibility to the general contractor in how it goes about awarding and administering subcontracts.

Chapter 8

Dispute Resolution

8.1 Contractual Jury Waivers Held Invalid By California Supreme Court
8.2 Subcontractor Forfeits Right To Arbitration By Filing Demand Untimely
8.3 Using Negotiation, Mediation And Arbitration To Resolve Construction Disputes.
8.4 Why Some Mediations Fail
8.5 Arbitration Consolidation Was Inappropriate
8.6 Architect's Decision Final Where Contractor Failed To Satisfy Arbitration Filing Requirements

8.1 Contractual Jury Waivers Held Invalid by California Supreme Court (Vol. 7, No. 8)

By: *Daven G. Lowhurst, Esq.*
Thelen Reid & Priest LLP

The Issue: The California Supreme Court has single-handedly altered the contractual expectations of parties to thousands of contracts governed by California law. For many years, parties to a wide variety of contracts – such as leases and construction contracts – have knowingly agreed to give up their right to have their disputes resolved by jury trial.

Indeed, these contracting parties likely continue to assume that any dispute that arises in the future will be subject to the jury waiver clause in their contracts. Confidence in the enforceability of contractual jury waivers could only have been increased beginning in 1991 when a California appellate court expressly validated such jury waivers.

That confidence was been shattered by the California Supreme Court recently in *Grafton Partners L.P. v. Superior Court*, 36 Cal.4th 944 (2005). The court ruled that contractual pre-dispute waivers of the right to trial by jury are unenforceable. Further, the Court elected to treat its ruling as *retroactive*, meaning that jury waivers are invalidated in existing contracts.

A brief trek through the Court's analysis is helpful to understanding how the Court reached its decision and what alternatives remain available to parties who wish to avoid the expense and uncertainty of entrusting their contractual disputes to juries.

The Decision: In *Grafton*, a partnership hired PriceWaterhouseCoopers (PWC) to audit accounts. Under the retention agreement, the parties agreed not to demand a trial by jury in any action arising out of PWC's services. The partnership sued PWC for negligence and misrepresentation arising out of the audit and demanded a jury trial. PWC, after unsuccessfully asking the trial court to strike the jury demand, obtained relief from the Court of Appeal. The Supreme Court reversed.

The court ruled that under California law, jury trials can be waived only when one of the six situations set out in Code of Civil Procedure §631 had occurred. Section 631 allows jury waivers when, for example, a party consents to waive jury in open court, fails to timely demand a jury trial or fails to pay jury fees. A contractual pre-dispute jury waiver is not one of those situations. Thus, the court found that such waivers violate the California Constitution's right to trial by jury.

In so ruling, the Court disapproved of the Court of Appeal decision in *Trizec Properties, Inc. v. Superior Court*, 229 Cal.App.3d 1616 (1991), which had validated pre-dispute contractual jury waivers by analogizing them to arbitration clauses. *Trizec* reasoned that if parties can contractually do away with trials altogether, they certainly can do away with trials by jury.

The Supreme Court found that analogy unpersuasive. First, the court explained that unlike jury waivers, arbitration clauses are expressly authorized by statute and advance a strong public policy in favor of arbitration. Second, arbitrations conserve judicial resources to a much greater extent than court trials by avoiding the judicial forum altogether. By contrast, court trials unquestionably tax the court system, even if to a lesser degree than jury trials.

The Supreme Court then ruled that its decision applies retroactively, *i.e.*, to pre-existing contracts. PWC argued that pre-dispute jury waivers are commonplace and that retroactive application of the court's decision would upset long-established contractual expectations. The court was not persuaded that an exception to the normal rule of retroactive application should be found because *Trizec* did not constitute a uniform body of law that could reasonably have been relied on in forming expectations.

A concurring opinion pleaded for the California Legislature to authorize pre-dispute jury waivers because they offer an "attractive middle ground" between jury trials and arbitration. Specifically, court trials calm contracting parties' fears of runaway jury awards while providing greater procedural and evidentiary safeguards than arbitration. Court trials also provide the additional safeguard of a full right to appeal rather than the very limited right to seek to vacate arbitration awards. Finally, court trials reduce the expense and delay associated with jury trials.

The Impact: *Grafton* broadly invalidates pre-dispute contractual jury waivers. What's more, it invalidates jury-waiver clauses in existing contracts, thereby shattering the contractual expectations of thousands of parties who have entered into contracts governed by California law.

Unless and until the California Legislature authorizes pre-dispute jury waivers, contracting parties seeking to avoid the expense and risk associated with jury trials may be limited to having their disputes resolved by arbitrators or by referees (usually retired trial court judges). The pros and cons of *arbitration* versus *reference* versus *jury trial* should be examined with legal counsel. Likewise, given that Grafton is retroactive, parties to existing contracts with jury waivers should consult legal counsel to determine whether to attempt to modify the contract to provide for arbitration or reference or whether to leave the invalid jury waiver as it is.

Whether in an existing contract or in a contract to be negotiated, if arbitration is preferred to having the dispute resolved by a referee, care must be taken to ensure that the arbitration clause satisfies California law, as different rules of arbitral enforceability govern different types of contracts. Further, when arbitration or reference is sought to resolve contractual disputes, care must be taken to ensure that the clause is not later invalidated as unconscionable. A pro-active approach to reviewing existing contracts with an eye toward dispute resolution will help ensure that any dispute is resolved in a manner consistent with the parties' expectations.

About the Author: *Daven G. Lowhurst, Esq* is an attorney with the law firm of Thelen Reid & Priest, LLP, located in San Francisco. For more information about the issues covered in this report, contact Mr. Lowhurst at 415-369-7270 or dglowhurst@thelenreid.com.

This article was originally published in the firm's newsletter and on its website (www.ConstructionWebLinks.com) in October 2005.

8.2 Subcontractor Forfeits Right to Arbitration by Filing Demand Untimely (Vol. 6, No. 4)

Where a subcontract required subcontractor to commence arbitration proceedings no later than 30 days following receipt of an adverse decision by the contractor, the failure of the subcontractor to contest a decision within 30 days was fatal to later seeking arbitration.

Holt, a drywall subcontractor, entered into a contract with Choate, the general contractor, to perform drywall installation on a new high-rise building. In the event Holt failed to meet its obligations under the schedule, the contractor was entitled to issue written decisions terminating its employment or supplementing its work with labor and materials the cost of which would be deducted from payment to Holt. The contract further provided that the subcontractor would be conclusively bound by, and abide by, contractor's decision unless the subcontractor timely commenced arbitration within 30 days following receipt of notice of a contractor decision, "otherwise, contractor's decision becomes final and binding."

On March 20 and again on April 12, the contractor notified the subcontractor that it was in default and must immediately remedy its defective performance or the contractor would hire additional or replacement contractors at the subcontractor's expense. On May 29 the contractor sent the sub a certified letter informing it of its decision to supplement the sub's work forces. The contractor thereafter employed additional workers and sent the sub change orders setting forth the amounts to be back-charged to the sub for this work.

It was not until November 5, after the contractor had closed out its prime contract with the project owner, that the sub filed a demand for arbitration. The contractor promptly filed suit in court to stay or stop the arbitration. The trial court agreed with the contractor and granted a stay to the arbitration. In affirming that decision on appeal, the appellate court explained that it found the plain language of the contract places the burden on the subcontractor to timely arbitrate any decision made by the contractor or be bound by it. The court was not impressed with the subcontractor's argument that the various notices and certified letter did not constitute a "decision" within the meaning of the contract. According to the court, the certified letter was a decision within the plain language of the contract.

The court was equally blunt in finding no merit to the subcontractor's assertion that the 30-day period to file an arbitration claim was impermissibly short. Relying upon applicable state law, and quoting case precedent, the court stated that parties to a contract may fix upon a shorter period for filing claims than that which is set by the state statute of limitations. Setting a shorter period violates no principle of public policy, said the court, "provided the period fixed not be so unreasonable as to show imposition or undue advantage in some way." For these reasons, the court held that because the subcontractor failed to avail itself of arbitration within 30 days, it was bound by the adverse contractor decision. *Holt & Holt, Inc. v. Choate Construction Company*, (2004 Ga. App. LEXIS 1602).

Comment: This case once again demonstrates the importance of knowing and abiding by the time requirements set forth in contracts. When commercial entities agree to contract terms and conditions setting forth various restrictions and limitations on submitting change orders requests, claims, and arbitration demands, the parties must be careful to abide by them. As seen by the decision in this case, unless there is a clear conflict with public policy, courts will not void the contract or re-write its terms.

8.3 Using Negotiation, Mediation and Arbitration to Resolve Construction Disputes. (Vol. 7, No. 3)

By: *Allan H. Goodman*

If a dispute arises during construction of your project, do *not* proceed immediately to court! Your contract may require you to use alternative dispute resolution (ADR) techniques such as negotiation, mediation or arbitration to resolve your disputes. You should therefore be familiar with these techniques.

Negotiation is the most informal method of dispute resolution. The only participants in the negotiation process are the parties to the contract and their designated negotiators. The goal of a negotiator is to resolve the dispute on the best terms for the party that he or she represents. When parties attempt to resolve a dispute by negotiation, it is not certain that the process will result in a resolution. The parties and their negotiators must deal face to face with each other in a manner that promotes the parties' interests, yet preserves the parties' relationship. In a successful negotiation, the parties and their negotiators reach a resolution of the dispute based on the parties' interests.

Mediation is often referred to as "assisted negotiation." In this process, the parties select a neutral person, the mediator, to help them arrive at a settlement of the dispute. A mediator is not a negotiator, as the mediator does not represent the interest of either party. The mediator is a facilitator, who helps the parties explore the strengths and weaknesses of their cases and assists them to frame and transmit settlement offers. The mediator spends a majority of the time meeting privately with each party. The mediator does not have authority to bind the parties, but can only help the parties resolve their dispute by agreement. In a successful mediation, the parties will reach a settlement of their dispute with the aid of the mediator.

As you can see, negotiation and mediation are both *non-binding processes* that will resolve the dispute only if the parties agree to a settlement. A settlement may be based upon compromises, promises of performance, and agreements to continue to do business in the future. By using these processes, the parties retain control of the resolution.

In contrast to negotiation and mediation, *arbitration* is *binding.* The parties select a neutral person, the arbitrator, who acts as a private judge. The arbitrator conducts a hearing, similar to a trial in court, and issues a decision, known as an "award," that binds the parties. Unlike a mediator, the arbitrator cannot meet privately with the parties, but must come to a decision based upon his or her understanding of the evidence submitted at the hearing and the law. Though arbitration is similar to litigation in court, it is private, the parties can choose an arbitrator that has particular expertise in the subject matter of the dispute, and the scheduling of the arbitration proceeding is not dependent on delays usually associated with a court's docket.

What if your contract does not require you to use ADR techniques or specifies a technique that the parties do not want to use? The parties may still use any ADR technique to resolve their dispute, as long as they agree. It is important that you know that litigation in court is not your only option.

About the Author: Allan H. Goodman is an experienced mediator and arbitrator of construction disputes, a Judge on the U.S. General Services Administration Board of Contract Appeals, and the author of *Basic Skill for the New Mediator* and *Basic Skills for the New Arbitrator*. He is also an instructor for Redvector.com, where he offers online courses on construction mediation and arbitration. For a detailed description of his books and courses, visit www.solomonpublications.com and www.redvector.com.

8.4 Why Some Mediations Fail (Vol. 7, No. 2)

By: *Allan H. Goodman*

Mediation is often used to resolve disputes arising during performance of construction contracts. Mediation is the non-binding, cooperative process in which parties to a dispute select a neutral third party, the mediator, to help them resolve their differences. The goal of mediation is a settlement agreement. While mediation is usually voluntary, many courts mandate an attempt to settle cases by mediation before setting a trial date. In a typical mediation, the mediator meets with all the parties and their attorneys in a joint session and then conducts a series of private, *ex parte* caucuses. In these private caucuses, the mediator assesses the strengths and weaknesses of the parties' positions and helps them frame settlement offers. The mediator may transmit settlement offers between the parties or may have the parties meet together in settlement discussions. A skilled mediator is able to help the parties overcome impasses in negotiations and guide the parties to a settlement.

While most mediations result in a settlement, some do not. A major cause of a failed mediation is that the participants approach mediation as informal, adversarial *litigation*. A party or attorney acting in this manner will address his or her remarks solely to the mediator, as if the mediator were a judge. Despite the mediator's efforts, the party refuses to have a dialogue with the opposing party and does not make a good faith effort to engage in settlement discussions. The mediation fails because the mediation process has never actually begun.

Another reason why a settlement may not be reached is that participation in mediation may reinforce a party's assessment that it has a very strong case and that a better result can be achieved in litigation. A party who has come to this conclusion has benefited from mediation without achieving a settlement. However, the party must proceed in litigation and achieve an acceptable outcome in order to validate its assessment.

Failure to achieve a settlement during mediation can also be caused by the personalities of the participants. Some disputes have their origin in or are sustained by personality conflicts that supersede the issues in dispute. If the parties and their attorneys treat each other disrespectfully before or during the mediation, this may cause psychological damage that prevents them from focusing upon and resolving the merits of the issues of the dispute.

A mediation may terminate without a settlement because of a participant's impatience. A party or its attorney may expect results too quickly or may think that the process is simply not working. Unrealistic deadlines or milestones for achieving results may be imposed on the mediator and the other party. This attitude is non-productive and self-defeating. A mediation conducted under such an arbitrary time frame will terminate simply because time has expired.

Finally, the parties may blame the mediator for the mediation's failure. There are good mediators and bad mediators, and the mediator selected by the parties or the court may not possess adequate skills to deal with the issues and personalities involved. On the other hand, dissatisfaction with the mediator may also result from the parties' unrealistic expectations or their own failure to commit to the process. The parties may fail to prepare for the mediation properly and refuse to attempt a dialogue with each other, but then demand that the mediator "do his magic." There is no magic in mediation. If the parties are not willing to prepare and work together toward a solution, it is difficult to achieve a settlement solely from the mediator's efforts.

About the Author: Allan H. Goodman is an experienced mediator and arbitrator, a Judge on the U.S. General Services Administration Board of Contract Appeals, and the author of *Basic Skill for the New Mediator* and *Basic Skills for the New Arbitrator*. He is also an instructor for Redvector.com, where he offers online courses on construction mediation and arbitration.

8.5 Arbitration Consolidation was Inappropriate
(Vol. 6, No. 7)

Where a subcontractor had a claim against the general contractor for payment, it submitted the matter to arbitration and the general contractor replied with a counterclaim in that arbitration and also initiated a third-party arbitration claim against a related subcontractor. The arbitrator ordered consolidation and permitted an indemnity claim between subcontractors. The impleaded subcontractor sought a court order to separate the arbitrations and to exclude the indemnity claim from the arbitration. The court denied that request. The subcontractor appealed and obtained a reversal, holding that the matters should not have been joined. Although each of the individual subcontract included a clause mandating arbitration, they did not dictate that matters involving separate subcontractors would be joined in a single arbitration proceeding.

In *Seretta Construction, Inc. v. Great American Insurance Co*, 869 So. 676 (2004), Pertree Constructors was the general contractor. It subcontracted with Seretta to erect concrete tilt-up walls, and it subcontracted with Five Arrows to prepare and paint the wall surfaces. The subcontracts included identical arbitration provisions stating: "Any claim, dispute or other matter in question between the Contractor and the Subcontractor relating to this Agreement shall be subject to arbitration upon the written demand of either party. Such arbitration shall be in accordance with the construction industry arbitration rules of the American Arbitration Association then obtaining. This Agreement so to arbitrate shall be specifically enforceable under the prevailing arbitration law, and the award rendered by the arbitrator shall be final, and judgment may be entered upon it in any court having jurisdiction thereof."

At issue in this case was whether it was appropriate for disputes under separate subcontracts to be joined into a single arbitration proceeding when each had its own arbitration provision and when joinder was contrary to the will of the subcontractors. In reviewing this issue, the court considered decisions of courts outside the state of Florida and found that some courts adopt the view that they have the power to direct that related arbitration matters be joined into a single arbitration proceeding. That line of cases relies upon doctrines of convenience and economy to resolve in one proceeding disputes which arise out of common facts and circumstances. In contrast to that line of cases, however, federal courts and several state courts hold that they have no power to compel consolidation where the contracts do not expressly provide for the same. They reason that for the court to "rewrite" the contracts or subcontracts to allow consolidation procedurally may conflict with the rights of one or more of the parties under their contracts.

The Florida court concluded that even though a consolidated arbitration in this case would be more expeditious and economical, the court would follow the approach of the federal courts and states that do not permit courts to order consolidation in the absence of an agreement by the parties to allow such consolidation.

Comment: As stated by the court in this case, it would have been more economical and efficient to combine the issues and disputes that arose out of common facts and circumstances into a single arbitration proceeding. Having a single arbitration enables a more prompt and logical resolution of responsibility for the various claims, counterclaims, damages, and indemnification between parties. Although courts in some states may impose consolidation despite the fact that the contracts do not call for such consolidation, this is certainly not always the case – as seen by the decision here. If you are signing an agreement containing arbitration provisions and you desire that all issues and matters related to common facts be resolved in a single consolidated arbitration, instead of piece-meal in different arbitrations, you should include express language in the contract to provide for such consolidation.

8.6 Architect's Decision Final where Contractor Failed to Satisfy Arbitration Filing Requirements (Vol. 5, No. 6)

Where a contractor failed to comply with arbitration notice and filing requirements, the architect's decision became final and binding, and the contractor had no further recourse to arbitrate or litigate its dispute with a homeowner.

The AIA form contract that was at issue in the case of *Martel v. Bulotti*, 65 P.3d 192 (2003), provided that in the event of a dispute, the parties were to submit the dispute to the architect for decision. Either party could demand arbitration after the architect submitted a written decision. Written notice was required to be filed by the party seeking arbitration with the other party to the Agreement, the American Arbitration Association [AAA], and the architect. The contract stated that failure to demand arbitration within thirty days would render the "Architect's decision ... final and binding."

In this case, a dispute arose concerning the contractor's performance. The parties submitted the dispute to the architect for decision, and the architect issued a written decision in favor of the homeowner against the contractor. The decision stated that it was final but subject to arbitration.

The contractor submitted a notice and demand for arbitration to the architect in the time permitted by the contract but failed to file it with the homeowner and the AAA as required by the contract. The homeowner then took the architect's award to court and applied for confirmation of that award and moved for summary judgment. In response, the contractor argued that he had substantially complied with the requirements for demanding arbitration and that the architect's decision was, therefore not final and binding.

The court found that the contractor failed to substantially comply with the terms of the contract because "substantial compliance" means that one party receives the important and essential benefits of the contract clause in question despite the deviation or omission by the other party. In this case, the court found that notice to the architect, (or even to the architect and the homeowner) would not trigger the arbitration process since it was not filed with the AAA. This denied the homeowner of an essential benefit of the contract, i.e., that disputes would be settled expeditiously and efficiently through arbitration with the AAA. This caused the architect's decision to become final and binding.

Comment: This case demonstrates the seriousness with which parties to a construction contract must take the terms and conditions addressing notice and filing requirements. In this particular case the contractor lost its right to contest an architect's decision because it failed to mail copies of the arbitration request to the homeowner and AAA when it faxed it to the architect. Numerous other decisions have been reported by courts in which a contractor is denied a change order because it failed to submit its request for a change within the period of time (e.g. 10 days) that is specified by the contract. Other cases have denied relief to contractors that submitted change order requests to individuals other than the individual that was specifically named in the contract as having authority to grant change orders. And this has been true even where there was no evidence that the project owner was harmed by the contractor's notice to the wrong individual.

There are also numerous cases holding that where a contractor proceeds to do changed work resulting from a differing site condition without giving prior notice to the architect (or in some cases the owner) as required by the contract, the contractor waives its right to recover its additional costs related to the changed work. Even where the parties have gotten into the habit, during the course of construction, of ignoring the niceties of notice and filing requirements, once a dispute ends up in court and attorneys get involved, the course of fair dealing and reasonableness between the parties often comes to a quick end as the attorneys read the contract documents and seek to strictly apply them to win their client's case.

Chapter 9

Documentation

9.1 Don't Touch That "Forward" Button! Attorney-Client Privilege in an E-Mail Age
9.2 Copyright Infringement of Design Documents

9.1 Don't Touch That "Forward" Button! Attorney-Client Privilege in an E-Mail Age
(Vol. 6, No. 6)

By: *Julie M. McGoldrick, Esq.*
Wickwire Gavin, P.C.

The attorney-client privilege is an important protection to enable businesses to seek open and frank legal advice in conducting their business affairs. When the privilege applies, statements and documents that normally would be open to inspection and discovery remain confidential, unless the client waives the privilege. The privilege enables clients to freely consult with their attorneys for confidential advice and can, therefore, help clients to avoid legal problems or disputes before they even arise. However, in the unfortunate situation that a dispute grows towards the need for a more formal resolution, privilege becomes even more crucial as it protects from disclosure documents that outline legal strategy or that explore strengths and weaknesses. In a time when e-mail communications are economical, convenient, and prevalent, it is particularly important to understand the basic principles of attorney-client privilege and to be aware of the unique problems that e-mail presents in protecting the privilege.

Because the attorney-client privilege is an exception to the general concept of open disclosure of evidence during litigation, the privilege is strictly construed. In order to be covered by the privilege, a communication must be made: 1) between an attorney and a client, 2) in confidence, and 3) for the purpose of seeking, obtaining, or providing legal assistance to the client.

Generally, the presence of or disclosure to a third party will prevent the privilege from attaching. There are some exceptions. For instance, if the third party is necessary to the attorney's complete representation of a client, the privilege may still attach. A necessary third party might be an accountant that the attorney has hired to help interpret books, or a paralegal researching an issue of law. Similarly, a client might require the presence of a translator in order to communicate with the attorney. Also exempt from the definition of "third party" are employees of a corporation that is represented by the attorney, as long as the employees are speaking about things within their scope of employment and understand that they are being questioned for the purpose of obtaining legal advice for the corporation. The general rule remains, however, that a third party who is not necessary to the representation will prevent the privilege from attaching.

Communications covered by the privilege remain confidential, unless the client waives the privilege. Once a communication is shown or repeated to a third party not covered by the privilege, the privilege is deemed to be waived. Waiver can occur voluntarily, such as when a client instructs the lawyer to reveal information to a third party (such as in settlement negotiations), or when the client herself reveals the communication. Waiver can also occur inadvertently. If, for instance, a client had a letter from her lawyer sitting out in plain view during a crowded meeting and a third party saw it, the privilege may be deemed waived as to that letter. A client must take reasonable steps to preserve the confidentiality of her privileged communications.

Just as e-mail has become incorporated into daily business activities, it is also a convenient and quick way for an attorney and client to communicate. E-mails may become privileged just like any other communications. Although the issue of the security of e-mails is beyond the scope of this article, the attorney-client privilege generally has been held to apply to e-mail communications. As such, e-mail is also subject to the same rules of waiver and, for the following reasons, e-mail communications are particularly susceptible to the unintentional waiver of privilege.

First, e-mail is easily shared. The "forward" button is an easy way to convey a lot of information with minimal effort. With the stroke of one key, the recipient can receive a string of e-mails that contain not only your instructions, but the history and context of the problem. However, if an e-mail from your attorney is included in that string, what once may have been a privileged communication may now be available for discovery during litigation if the recipient of the e-mail string is a third party. An e-mail from your attorney is privileged, but once you forward it to anyone not covered by the privilege, the e-mail is no longer a confidential communication.

Inadvertent waiver is particularly a risk with e-mail software containing an auto-text feature that automatically completes e-mail addresses. This convenient feature is perilous to the attorney-client privilege, especially if you have more than one contact with similar e-mail addresses. You may think you are sending your estimation of damages to John White, your attorney's paralegal, but in fact, your computer took the liberty of sending it to the first "John" in your recent directory—John Smith, the subcontractor you are considering suing.

Just being aware of the relatively frail nature of the privilege in e-mail can go a long way towards protecting your confidential communications. There are also a few precautions to take in order to lessen the chance of inadvertently waiving privilege:

- As noted above, take care not to forward e-mails from your attorney. If your attorney e-mails you with advice about how to handle a situation, start a new e-mail to give instructions to the proper people.

- Similarly, when following your attorney's advice, it is usually better not to explicitly state that. Instead of writing, "My attorney said that I should research the issue of my liability for negligence on the job before I sign anything," just request the information that you need.

- Remember facts are not privileged, so you can share factual information with anyone without worrying about waiving your privilege. It is your attorney's advice and counsel that is protected. If you reveal that, you may be inadvertently waiving the privilege over that communication.

If forwarding e-mails is a convenience you cannot forego, at least be sure to forward only the e-mail that is relevant to the recipient. That way, you are less likely to inadvertently send privileged information. Likewise, be particularly careful if your e-mail software employs an auto-text feature for e-mail addresses. Consider deactivating this feature if possible and double check the recipient list before you send out a particularly sensitive message.

Finally, if you want to communicate information and you are worried that it might waive your privilege, it is always a good idea to consult your attorney. She may be able to communicate factual information orally for you in such a manner as to avoid concerns of waiver.

About the Author: Julie M. McGoldrick is an attorney with the Los Angeles office of Wickwire Gavin, P.C., and focuses her practice on construction law matters. She may be reached by telephone at 213-688-9500 or by email at jmcgoldrick@la.wickwire.com..

9.2 Copyright Infringement of Design Documents

(IRMI Expert Commentary, K. Holland - November 2002)

Instruments of service produced by the design professional, including plans, specifications, drawings, opinions, reports, and calculations have historically been treated as intellectual property belonging to the design firm that created it. This has been plainly stated in standard form contracts such as those published by the American Institute of Architects (AIA), in Document B141, and the Engineers Joint Contract Documents Committee (EJCDC) in EJCDC Document 1910-1. This article examines the ownership of such documents and examines a recent copyright case over an architect's drawings.

Ownership of Documents

A sample contract clause protecting the design firm's ownership rights is as follows:

> "Drawings, specifications and other documents, prepared by the Design Professional (DP) and the DP's consultants

> are Instruments of Service for use solely with respect to this Project. This includes documents in electronic form. The DP and the DP's consultants shall be deemed the authors and owners of their respective Instruments of Service and shall retain all common law, statutory and other reserved rights, including copyrights. The Instruments of Service shall not be used by the owner for future additions or alterations to this Project or for other projects, without the prior written agreement of the DP. Any unauthorized use of the Instruments of Service shall be at the Owner's sole risk and without liability to the DP and the DP's consultants."

The ownership clause like the above-quoted one sets forth clearly the rights of the design professional and protects against the risk of liability that might otherwise arise out of reuse of the documents by an unauthorized person, including the project owner. The protection afforded by this clause is appropriate because if the documents are used on other projects without the design firms knowledge and input, the designer will be unable to assess and revise the design for the new circumstances or new project on which they are being utilized. This means he or she will not be able to manage the risks that will naturally arise when design documents are used on a project.

In contrast to the reasonableness of the AIA and EJCDC clauses, the provision of the contract form of at least one owner organization states the following:

> The Construction Documents and any other documents or electronic media prepared by or on behalf of the Professional for the Project are the sole property of the Owner free of any retention rights of the Professional. The Professional hereby

unconditionally transfers and assigns to the Owner all copyright claims, trade secrets or other proprietary rights with respect to such documents, and agrees, upon request of the Owner, to turn over to the Owner the originals and all copies of such documents and materials as of the date of such request.

Indemnity Clause

If an owner is insistent that it be given ownership rights to the design documents, and you decide as a matter of business judgment that you are willing to grant such rights, you should seek to add an indemnity clause to protect you against claims that might arise out of the reuse of the documents. For example, you might include language like the following.

The Owner agrees to hold harmless, indemnify, and defend the design professional against all damages, claims, and losses of any kind (including defense costs), arising out of any use of the plans and specifications on any other project, for additions to his project, or for completion of this project.

You should also be careful not to give away your own right to reuse the documents in the course of your future services for other clients. The EJCDC Document 1910-1 (clause 6.04) handles this by stating: "Engineer shall retain an ownership and property interest therein (including the right to reuse at the discretion of the Engineer) whether or not the Project is completed."

Potential Liability Exposures

The problem with allowing the owner to reuse your documents, besides the obvious fact that you are giving your work away for free, is that you lose control over how the documents are interpreted and used. This puts you at significant risk since you will not be able to make necessary revisions and changes to the documents that may be necessary before the can be used successfully on the new project. The liability exposure from such reuse should be carefully considered before you agree to permit it, and before agreeing to permit such reuse, it is advisable to negotiate specific disclaimers on the reuse and indemnification from the owner.

Recent Case: There have been a number of cases in which an architect's drawings were used to complete a project by a different architect when the original project developer transferred the project to a new developer or design-builder. In several of these situations, the original architect successfully sued the new developer for the unauthorized use of his design documents. A recent example is the case of *Nelson-Salabes v Morningside Development*, 284 F3d 505 (4th Cir 2002). In that case, the original architect ("NSI") performed professional services for the original developer ("Strutt") in three separate phases. In the first phase, NSI delivered to Strutt a proposed letter agreement under which NSI agreed to develop a schematic building footprint for an assisted living center called Satyr Hill. Although Strutt never executed the agreement, both Strutt and NSI fully performed according to its terms.

Next, NSI submitted a proposed letter agreement to provide additional architectural services to develop the exterior elevations for the project and attend a zoning exception hearing. Again, all terms of this proposed agreement were performed by Strutt and NSI although Strutt never actually signed the agreement. After this, NSI created four architectural drawings depicting the building footprint, the floor plans, and the exterior elevations. These were incorporated by Strutt's civil engineer into the development plan for the project and submitted to the zoning board which granted the request for a zoning exception.

While the zoning application was pending, NSI submitted a third proposed letter agreement to Strutt offering to create the design and working drawings for the remaining development of the project. This proposal stated, "If the above is acceptable, we will prepare a Standard AIA Agreement." Consistent with its record, Strutt did not execute the letter agreement. Several months later, NSI submitted a revision to this proposed agreement along with a "revised AIA Contract for Satyr Hill Catered Living per our recent discussions." The AIA Contract provided in relevant part that "[t]he Architect's Drawings, Specifications or other documents shall not be used by the Owner or others on other projects, for additions to this Project, or for completion of this Project by others unless the Architect is adjudged to be in default under this Agreement, except by agreement in writing and with appropriate compensation to the Architect." Once again, Strutt failed to sign this agreement. One month later, Strutt advised NSI to cease performing services because Strutt's potential business partner had backed out of the project, and Strutt lacked sufficient expertise to go forward with the project alone.

In an interesting twist, Strutt asked NSI if it might know of any potential buyers of the project that could complete it. NSI then successfully solicited buyers on behalf of Strutt and as a result a group called "Morningside Development" took over. Ironically, however, Morningside decided to consider different architects to complete the project. NSI advised Morningside that if it did so it had no authority to use the NSI drawings without its express written consent. Morningside thereafter entered into a design-build contract for construction of the project and provided the design-builder ("Hamil Commercial") with a copy of the NSI drawings. The design-builder in turn gave the drawings to its subcontracted architect ("EDG Architects"). Morningside then met with EDG and instructed it to avoid any modifications to the original plans and drawings that would necessitate obtaining a new zoning exception. After the project was completed,

NSI Architects filed suit against Morningside alleging copyright infringement for unauthorized use of NSI's design documents.

In their defense, the defendants argued that they could not be held liable because they had an "implied nonexclusive license" to use the NSI drawings. They argued that the totality of NSI's conduct implied the existence of such a license. In analyzing whether such an implied license had been created, the court concluded that an implied license is created when three conditions are met, including "(1) a person (licensee) requests the creation of a work, (2) the creator (licensor) makes that particular work and delivers it to the licensee who requested it, and (3) the licensor intended that the licensee copy and distribute the work."

The third element of this test was not met in this case, said the court, because NSI did not intend that its copyrighted drawings be used on the project for which they were created independent of NSI's continued involvement. Nothing about NSI's representations or conduct suggested that it intended either the original developer or Morningside to use its plans without NSI's future involvement or express consent. In fact, NSI specifically advised Strutt to the contrary. The court made particular note of the fact that NSI submitted an AIA agreement to Strutt that stated NSI's intention that its drawings not be further used without its express consent. For these reasons, the court held that NSI did not grant a implied license to the defendants to use its drawings.

Comment:

Several lessons are learned from this case. It demonstrates the importance of using agreement forms, such as those of the Design-Build Institute of America (DBIA) or the American Institute of Architects (AIA), that preserve the copyright interest of the architect. It demonstrates the importance of getting things in writing but shows that even when written agreements are not signed, the actions of the parties in performing in a manner consistent with the terms and conditions of the unsigned contract may be evidence of the contractual intent of the parties. Another issue is the importance of choosing clients that are financially sound and have experience with similar projects and contracts so that expectations may be managed and the project may be completed as anticipated by the design professional. Finally, it is somewhat surprising that the architect here apparently did not obtain any written assurances from Strutt before it assisted Strutt in finding another developer to buy the project, and that it likewise did not obtain any written assurances of the new developer, Morningside, before introducing it to the project.

Chapter 10

Drug Testing

10.1 Rapid Result Drug Testing (Vol. 6, No. 2)

By: *William F. Current*
WFC & Associates

Would you be surprised to learn that nearly a quarter of your workers were illegal drug users? It could happen. Consider what a California-based contractor discovered several years ago when it decided to prove that its workers were *not* druggies.

On a given day following 30 days advance notice employees were asked to volunteer for a drug test. Of the 179 people on the payroll, 80 volunteered. The urine samples were collected by an independent laboratory, tested and reported directly to the contractor. No names were used in either the collection or the reporting of the results. Three different construction sites were chosen to represent a cross section of employees from northern, central and southern California.

The results speak for themselves: the percentage of tests that were positive for one or more drugs was 24 percent, a quarter of the employee population. From that group 15 percent tested positive for marijuana and 10 percent tested positive for cocaine.[i] Wow! And that was from a group of volunteers.

The Drug Problem Today

America has made progress in addressing its drug problem over the past 10 years, but unfortunately the problem has not gone away. The federal government's annual report on substance abuse indicates just how serious the issue is.

For example, there are approximately 19 million current (use in the last 30 days) *illicit* drug users 18 and older. Further, there are about 35 million *prescription* drug abusers. There are nearly 16 million adults 21 and older who admit to being "heavy" drinkers (5 or more drinks on at least 5 or more occasions every month); and there are 2.3 million Americans younger than 21 who admit to being heavy drinkers.[ii]

Several national reports on teen substance abuse do not paint a promising picture of the near future outlook. And, of course, today's teen drug user is tomorrow's job applicant.

Drugs in the Construction Workplace

The federal government estimates that 77 percent of all illicit drug users 18 and older are employed.[iii] And the industry that is often identified for having the highest rate of illicit drug users is the construction industry.

A 1996 federal government survey, the last such survey conducted by the government, showed how serious the problem is for construction companies. Among full-time construction workers between the ages of 18 and 49 more than 12 percent reported illicit drug use in the month before the survey was conducted; almost 21 percent reported illicit drug use during the past year. Additionally, approximately 13 percent admitted to being "heavy" alcohol users.[iv]

Rates of substance abuse among different occupations in the construction industry included:

Position	Current Illicit Drug Use (%)	Past Year Illicit Drug Use (%)	Current Heavy Alcohol Use (%)
Construction Laborers	12.8	25.4	19.9
Construction Supervisors	17.2	25.9	12.7
Other Construction Workers	17.3	23.4	20.6

How Drug Abuse Affects the Construction Industry

Generally speaking, we know that substance abusing workers are less productive, tend be unreliable, are more likely to be involved in workplace confrontations and acts of violence, and steal from their employers and others at a higher rate than their non-using co-workers.

A compelling study by the U.S. Postal Service found that substance abusers, again when compared to their non-substance abusing co-workers, are involved in 55 percent more accidents, and sustain 85 percent more on-the-job injuries.[v] Further, the National Safety Council reported that 80 percent of those injured in "serious" drug-related accidents at work are not the drug abusing employees but innocent co-workers and others.[vi]

It also stands to reason that if the construction industry employs more drug users than other industries, then the impact of drug abuse would be significant at construction sites. And, given the safety sensitive nature of the construction industry, that impact is most significant in the area of safety.

A study conducted by a Cornell University graduate student found that construction laborers between 25 and 34 years of age who have been treated for substance abuse have a time-loss injury rate of 23.6 per 100 full-time equivalent workers (FTEs). That's nearly double the rate of non-substance abusers, who had a rate of 12.2 FTEs.

The majority of the cases, 85 percent, involved treatment for alcohol abuse. The study concluded that the difference between the known substance abusers and the non-substance abusers is "likely understated." The study observed that:

"Injuries were counted as related to substance abuse only after substance abuse was diagnosed, yet 1/3 of the substance abusers' work-related injuries occurred before diagnosis."[vii] The study only tracked workers on union jobs and only substance abusers treated in programs paid for by union health insurance.

Drug Testing As a Solution
The construction industry is especially affected by drug abuse given that it tends to hire a higher proportion of substance abusers. As a result, employers in the construction industry tend to be very concerned about the issue and are probably more likely to have a comprehensive drug-free workplace program in place. And among the components of such a program is drug and alcohol testing.

Drug and alcohol testing have proven to be highly effective ways of deterring substance abuse and identifying those who need help. It is legal in every state, though a handful of states regulate it, and commonly accepted as way of life in American industry.

For many years construction companies that drug test have utilized the services of a laboratory certified by the federal government's Substance Abuse and Mental Health Services Administration (SAMHSA) to analyze all drug screens. And for many years this was the best testing method available. However, while drug testing makes a lot of sense for many construction companies waiting 2-4 days to get a result from a lab is highly impractical. When a drug test result is the only thing stopping a crew of 20 or 30 new workers from starting a job, an immediate, accurate result is really what is needed.

Rapid result, on-site testing has become a popular alternative to traditional lab-based testing, especially in safety-sensitive industries such as the construction industry. Often a construction company can significantly reduce the time it takes to conduct a test by utilizing rapid result testing. The results are available within minutes and, depending on the product being used, can be as accurate as the screening technologies used in laboratories.

When accidents or some other unacceptable behavior occurs, construction companies rarely have the luxury of waiting for a lab result to come in 24-48 hours later. Rapid result testing is a viable option for post-accident and reasonable suspicion drug testing.

Rapid Drug Testing Is Union Friendly
Union members are typically in favor of drug testing. A Gallup survey found that 71 percent of full-time union workers favor employers' right to conduct pre-employment testing. Further, when asked, "would you favor or oppose *your* company adopting or maintaining a drug testing policy," 66 percent of union respondents said they would favor such a policy compared to 26 percent who said they would not.

Union workers are just like any other workers, the vast majority are not drug users, yet they know who the drug users at work are and they don't like working side by side with them, especially in safety-sensitive worksites.

The general concerns that some union may have about drug testing are addressed with rapid drug testing. Rapid drug testing delivers fast results making it possible for workers who test negative to get back on the job quickly.

Unions are interested in the integrity of the testing process. A rapid drug test can be witnessed throughout every phase of the testing process. Workers actually get to see the entire test take place. There is never a question about chain of custody. And because the analysis can be witnessed there is rarely any confrontation between the tested worker and the test administrator over a result.

Unions are interested in preserving the clean records of their members. Again, because the majority of all drug screens are negative, tested workers are not only back on the job faster, but there's no lingering question about the result as management, the union, and workers wait a couple of days for a lab-tested result to come back.

Oral Fluid Testing
Construction employers also now have options available in terms of what specimen to test. No longer is urine the only specimen recognized as an accurate medium for detecting drugs. Oral fluid and hair samples have proven to be effective in detecting drugs of abuse. Oral fluid testing, in particular, is an attractive option to the construction industry because it can be conducted either with a rapid result device or through a laboratory.

Oral fluid testing eliminates the inconvenience of securing a restroom in the middle of a construction site. It makes it possible for every collection to be observed, and eliminates concerns about mixed gender collections.

Studies show that oral fluid testing is an accurate indicator of the presence of drugs in a person's system.

When Choosing a Rapid Result Testing Device

A word of caution about rapid result testing: Not all of these testing devices are created equal. While the prices of these products have come down significantly in the last few years, the old adage "you get what you pay for" often comes into play. When considering instant testing, be it with urine or oral fluid, consider the following:

1. Make sure your state allows it. Most states have no restrictions on instant testing or oral fluid testing, some do.
2. Look for urine devices that are approved by the U.S. Food & Drug Administration (FDA). This is quickly becoming the gold standard for instant urine testing. Instant oral fluid testing has not passed FDA muster yet, though are there several reliable products available.
3. Limit your considerations to devices that come with independent scientific data to back up all accuracy claims. It's not enough to see it in writing; make sure the source is an independent, objective one.
4. Try all devices under consideration in real work situations. Devices differ in how they are administered, how results are read, how long it takes to get a result, etc. Make sure the device you're considering will work for you.
5. Deal with a provider who has been in the business more than a few months (and maybe even years). You're going to need support, both technical and perhaps legal. Not all providers have staffs of experts on hand to answer your questions.

About the Author: Mr. Current is a principal with the firm WFC & Associates, Substance Abuse Prevention Consultants, located at 8627 N.W. 50 Drive; Coral Springs,

FL 33067; Ph: 954-255-8650; Fax: 954-344-0707. His e-mail is: bcurrent2@aol.com.

---

[i] California Associated General Contractors, 2005.
[ii] National Survey of Drug Use and Health. Substance Abuse and Mental Health Services Administration. Washington , D.C. 2003.
[iii] Ibid.
[iv] U.S. Department of Labor. Substance Abuse Information Database (SAID) website.
[v] Current, William F. Why Drug Testing? Coral Springs , FL. 1999.
[vi] Ibid.
[vii] Meyers, Linda. "Construction Company Drug Testing Reduces Work Injuries, Study Finds." Cornell University . 2000.

Chapter 11

Economic Loss Doctrine

11.1 Contractors May Now Bring Direct Action for Economic Losses Against Design Professionals in Pennsylvania (Vol. 6, No. 2)

By: *Andrew B. Cohn, Esq.*
Kaplin Stewart Meloff Reiter & Stein, PC

A very recent Pennsylvania Supreme Court opinion (January 2005) has significantly changed Pennsylvania law, allowing a general contractor to directly sue an architect in negligence for additional construction costs caused by defective plans, drawings, and specifications. Based on this decision, contractors working on Pennsylvania projects may now bring direct actions against design professionals to recover purely economic losses caused by errors and omissions in their design documents.

In *Bilt-Rite Contractors, Inc. v. The Architectural Studio,* a general contractor on a public school project claimed that it incurred substantial additional construction costs because the aluminum curtain wall, sloped glazing, and metal support systems could not be installed and constructed through the use of normal and reasonable construction means and methods. The contractor claimed the Project drawings and specifications were deficient and caused the additional costs, which resulted from special construction means and methods which the contractor was compelled to utilize.

The general contractor sued the architect directly under the theory of "Negligent Misrepresentation", alleging that the architect's specifications were false and/or misleading. The contractor had contracted with a School District , and therefore it did not have a direct contract with the architect. Moreover, its damages were purely *economic* (i.e., it did not allege bodily injury or physical damage to property). The trial court dismissed the contractor's suit, and the Pennsylvania Superior Court affirmed the decision, applying established Pennsylvania law which had previously held that a third party contractor could not directly sue a design professional for negligence causing purely economic losses.
However, the Pennsylvania Supreme Court reversed, holding in a decision of first impression by the Supreme Court, that a contractor can sue a design professional in negligence for purely economic losses resulting from defective plans and specifications.

The Supreme Court reasoned that contractors fall squarely within the class of companies which reasonably rely on the representations of design professionals in their design documents. Applying Section 552 of the Restatement of Torts (2d), the Court stated that it was reasonable for a design professional to expect that contractors would rely on information supplied in project design documents, and reasonably foreseeable that contractors could incur economic losses if the design information was incorrect or erroneous. These expectations, according to the Court, reflected modern business realities which justified holding design professionals responsible for economic harm caused to those who rely on this information in project plans and specifications.

The significance of this decision cannot be overstated. In addition to their contractual responsibility to their clients,

Pennsylvania design professionals will now be exposed to direct causes of action for economic losses sustained by contractors who rely on defective plans and specifications. Such damages may exceed those sustained by a design professional's clients (e.g. project owners). On the other hand, contractors (and most likely subcontractors) now have the option, in addition to seeking change orders from owners under their contracts, to bring a direct action against a design professional for increased construction costs caused by design errors and omissions.

About the Author: **Andrew Cohn** is an attorney with the law firm of Kaplin Stewart Meloff Reiter & Stein, PC, 350 Sentry Parkway Building 640, Blue Bell, PA 19422; acohn@kaplaw.com; 610-260-6000.

Chapter 12

Environmental Liability

12.1 Superfund Decision May Benefit Design Professionals on Environmental Remediation Projects

(IRMI Expert Commentary, K. Holland April 2001)

Design professionals and contractors have long been concerned about their own potential liability under the environmental liability of the Comprehensive Environmental Response, Compensation, and Liability Act (CERCLA or "Superfund"). This article discusses the decision of *Bashland, Bouck & Lee v City of North Miami*, 96 F. Supp. 2d 1375 (SD Fla 2000), and its affect on such liability under Superfund.

The Problem

Numerous courts have imposed strict liability on design professionals and contractors for environmental impairment resulting from their services. In one case, a contractor was held to be subject to strict liability as a "transporter" under Superfund for moving contaminated backfill around a construction site. This was despite the fact that the contractor did not know the material was contaminated and did not move it off-site.

In another case, an engineer was held to be subject to strict liability as an "arranger" for disposal of hazardous waste on the basis of having provided a site owner with the names of several licensed disposal facilities as options for disposing of wastes removed from the site during cleanup activities.

Although several trade associations frequently lobbied Congress to amend the statute to protect contractors and design professionals from being sued for strict liability under Superfund as a "potentially responsible party" ("PRP") on projects for which they performed services, no meaningful relief has ever been enacted. In managing their environmental remediation contracts and projects in a manner to reduce the risk of such liability, contractors have sought to place ultimate responsibility for decisions concerning disposal site selection, transportation, and operations into the hands of others, such as the site owner or the design professional. This risk management device may be less effective or even impossible, however, when the remediation is performed as a design-build project.

In the context of design-build projects, there has been concern that the team of the design professional and contractor would have a greater potential liability under CERCLA because together they are responsible for making many of the decisions affecting the design, testing, and movement of soil and possibly groundwater that could have environmental impact. The design-builder is more likely to be accused by a site owner or government agency of having PRP liability because it is the single point of accountability for decisions affecting the environment.

The Case

In *Bashland, Bouck & Lee v City of North Miami*, the City hired an engineering firm ("BB&L") to assist it in implementing an environmental cleanup plan for a landfill that was on the U.S. Environmental Protection Agency (EPA) National Priorities List because of uncontrolled hazardous releases that included ammonia leaching into the groundwater. The firm conducted hydrogeologic studies of the aquifer and used those studies to design a leachate collection system. It also did pump tests, computer modeling, surface and groundwater sampling, and design services related to relocating an earthen dike.

The City terminated the engineering contract and refused to pay the balance of fee that was due. BB&L sued the City to recover its unpaid fee. In its defense, the City argued that the engineer was a "potentially responsible party" (PRP) and, therefore, barred from recovering the costs it claimed. The crux of the City's argument was that the engineer allegedly contributed to increased cleanup costs and had liability under CERCLA as either an "operator" of a hazardous waste facility or an "arranger" for the disposal of hazardous wastes at a facility.

In reviewing the matter, the court concluded that negligent performance of services by the engineer would not subject the engineer to PRP liability. As explained by the court, a PRP is an owner or operator of a CERCLA facility. The basis for the City's assertion that the engineer was an "operator" or "arranger" included the following theories:

- The engineer was an "operator" or "arranger" because it had "operational control" of the site;
- The engineer failed to develop an adequate remedial design which permitted the ammonia to continue migrating through the groundwater; and

- The engineer's excavation of soil in order to obtain test samples of the waste that was already in the ground contaminated the "clean cover" and, therefore, constituted "arranging" disposal of hazardous waste.

All three theories were rejected by the court, which concluded that "engaging in cleanup activity at a facility does not qualify as the type of 'operation' CERCLA contemplates." Before a party can be found to be an "arranger," the court said that the "party must take an affirmative step to introduce hazardous substances to an area—mere inaction or inept action which fails to remedy but does not worsen existing contamination is not sufficient."

The City did not allege that the engineer brought any contaminant onto the site and released it. "Here, the City alleges BB&L's negligence enabled ammonia to continue to migrate through the landfill. This is insufficient to establish arranger liability because no affirmative step was taken."

The court also rejected the City's argument that BB&L's action in refilling excavated pits with the same soil that was removed for testing purposes constituted "disposal." This gets to the question of whether the mere act of inadvertently moving contaminated material around a site is enough to render a firm liable under the strict liability aspects of CERCLA. According to the court, it does not.

A second flaw with the City's argument, according to the court, was that "a response action contractor cannot be liable for response costs unless it is negligent and such negligence causes the release of a hazardous substance." In this case, although there had been negligence in the remedial design services, the court found that there was no evidence that the soil investigation had been performed negligently. The court concluded that the engineer's excavation activities did not cause it to become a CERCLA arranger, and that it was, therefore, not a PRP within the meaning of the law.

Comment

By subjecting designers and contractors to Superfund strict liability in cases such as that reported here, other courts have caused a chilling effect on the ability and willingness of designers and contractors to perform work related to the cleanup of contaminated sites. It is to be hoped that the logic of this decision will be followed by other courts around the country. The decision of this court sets forth some very practical principles on which to defend against such liability.

Chapter 13

Ethics

13.1 Testing Your Ethical Barometer (Vol. 5, No. 6)

By: *Michael Loulakis, Esq.*
Wickwire Gavin, P.C.

The barrage of corporate misdeeds reported over the past year makes one wonder how things could have possibly gotten so out-of-hand. How could Enron's management get away with "cooking the books" and reporting phantom revenue so easily for so long? Why didn't someone at Arthur Andersen look at the big picture and figure out that bad things (like going out of business!) would happen if they caved into the pressures of satisfying a big client?

Many of us in the design-build community look at these events and shrug them off as shenanigans played by "the big boys" on Wall Street, where the term "business ethics" is an oxymoron. After all, design-build has gained market share over the past decade because of what it sells – professionalism, honesty and teamwork. But is it fair to say that design-builders possess only the purest of business ethics? Probably not.

Consider how design-build is marketed. It is easy to sell an owner about the concept of single point responsibility – if something goes wrong, the design-builder will take care of it. But is this really what the design-builder means? Are contractor-led design-builders really willing to offer the owner more protection than what the contractor can get from their design teammates or insurance? Will design-led design-builders agree to warrant that their design will achieve performance guarantees in the contract? Will they pick up responsibility to the owner if the contractor fails to perform?

What about the process of selecting the design-builder and its team? The Design-Build Institute of America (DBIA) and design-builders extol the virtues of qualifications-based selection (QBS) for design-build services. Yet how many design-builders practice what they preach and select subcontractors and professional consultants on the basis of QBS? The reality is that the design-builder's team is often selected through the same low price / bid shopping mentality that has led to so many problems under other delivery systems. Is this in the best interests of the project?

Consider project execution. Design-builders often follow the same pattern as low bid general contractors in preparing and updating project schedules. The design-builder may keep three different schedules on a project – one for the owner, one for the subcontractors, and the "real" schedule. Schedule graphics can be easily manipulated to hide logic ties and show activities as "critical" when they are not. How many design-builders are willing to provide owners with copies of their electronic schedules and allow the owner to see the true picture of the job?

What happens if the owner is a tough negotiator? Is it wrong to add a little bit extra here and there in a change order request to give some bargaining room? And if you know that you are going to have a dispute, do you create a few claims so that you can horse-trade later? Is this fair, open and honest?

It is likely that the management teams of Enron viewed their conduct as being within the rules of the game. I suspect that some design-builders operating in today's environment feel the same way. Those working in the public arena may not have heard about statutes like the Truth in Negotiations Act and the False Claims Act. Violations of these statutes carry heft fines and potential debarment. Do design-builders working in the private sector every think about the prospect of being sued for fraud or under a state deceptive trade practices statute? Do they ever think that what happened to Arthur Andersen in terms of loss of reputation and confidence can happen to them?

It may be impossible to test an industry's ethical barometer. My own experience tells me that most in the design-build community go out of their way to ensure that they are being fair, open and honest with their counterparts. What does your ethical barometer tell your colleagues about you?

About the Author: **Michael Loulakis** is President of Wickwire Gavin, P.C., and a shareholder in its Virginia office, located at 8100 Boone Blvd., Vienna, VA 22182 . He devotes his legal practice to representing parties in the construction industry, including owners/developers, sureties, contractors, and design professionals. He can be reached at 703-790-8750 or by e-mail at mloulakis@wickwire.com. This article is reprinted from the February 2003 issue of Design-Build DATELINE (Design-Build Institute of America).

Chapter 14

Expert Witnesses

14.1 Contractor Complaint against Engineer Dismissed for Failure to File Expert Identification Affidavit
14.2 Personal Injury Case against Engineer Dismissed for Lack of Expert Testimony

14.1 Contractor Complaint against Engineer Dismissed for Failure to File Expert Identification Affidavit (Vol. 6, No. 6)

Where a contractor that was sued by a project owner for failing to comply with contract specifications filed a claim to implead the project engineer into the suit, it failed to serve an expert identification affidavit within the following 180 days as required by state statute. As a result, the court dismissed the contractor's claim against the engineer.

In *Middle River-Snake River Watershed District v. Dennis Drewes, Inc.*, 692 N.W.2d 87 (2005 Minn.), a contractor, Dennis Drewes, Inc. was contracted by a watershed district to work on a flood impoundment project. Before submitting its bid, the contractor reviewed soil reports that indicated that the soil would be ideal for construction. The contract required the contractor to achieve an overall soil compaction of 95 percent and prohibited lifts greater than twelve inches in height. During construction, the contractor encountered wetter soil conditions than anticipated but instead of notifying the district of the changed conditions the contractor deviated from the specifications and used lifts greater than twelve inches in order to complete its work and failed to meet the compaction requirements.

Upon learning of the contractor's non-compliance the district filed suit against it. The contractor responded with counterclaims against the district and also filed a claim against the project engineer, J.O.R. Engineering, Inc., alleging negligence, estoppel, and tortuous interference with its contract. Contractor failed to comply with the expert witness disclosure requirements of the Minnesota statute. The statue require service of two affidavits on the adverse party. The first is an affidavit of expert review to be filed with the pleadings. The second is an expert-identification affidavit that must be served within 180 days after the first affidavit. A party's failure to provide the second affidavit "results, upon motion, in mandatory dismissal of each action with prejudice as to which expert testimony is necessary to establish a prima facie case."

In this case, the contractor filed the expert-review affidavit but failed to serve the second affidavit within the required 180 days. The engineer then filed a motion to dismiss the complaint for failure to file the affidavit. In response to that motion, the contractor, three days later, filed the missing affidavit.

Under the state statute, there is a 60 day cure period for a defective affidavit. It states that a motion to dismiss may not be granted "unless, after notice by the court, the nonmoving party is given 60 days to satisfy the disclosure requirements." The contractor argued that since it filed the affidavit within 60 days of the motion to dismiss, it is exempted from the mandatory dismissal provision. In rejecting this argument, the court held that the 60 safe harbor period is only for "claimed deficiencies of the affidavit" and does not apply in the situation of a complete failure to file any affidavit (albeit defective) within the mandatory 180 day period. The court said that this was self evident from the fact that the last sentence of this section of the statute requires the court to issue specific findings on "the deficiencies of the affidavit." As explained by the court, "When an initial expert-identification affidavit has not been filed, a court would be unable to make specific findings on the deficiencies." For these reasons, the court affirmed the summary judgment against the contractor.

Comment: Numerous states have statutes requiring some type of expert affidavits to be filed with pleadings and/or to be served on the other party within some period of time following the pleadings. Failure to timely file these affidavits has been the basis for a number of judicial decisions dismissing actions against professionals. This current decision is a reminder of the importance of knowing and following in careful detail the requirements of statutory requirements concerning what must be filed and when it must be filed. Since dismissal with prejudice is mandatory under these statutes, the courts had little leeway (or even no leeway) to do anything but grant a motion to dismiss.

14.2 Personal Injury Case against Engineer Dismissed for Lack of Expert Testimony (Vol. 6, No. 5)

Where a pedestrian sued a city and its engineering consultant for negligent design and construction, and failure to warn of a dangerous condition in a sidewalk, a court held that the engineer was entitled to summary judgment. This was because the plaintiff did not present expert testimony on the professional standard of care, and the evidence did not establish a duty of the engineer to warn of a dangerous condition.

In the case of *Luther v. City of Winner and Dan Britton*, 674 N.W. 2d 339, 2003 WL 23137692 (South Dakota, 2004), the renovations for Main Street that was designed by the city's engineer included changing the sidewalk and curbs to correct drainage problems caused by the fact that one side of the street was two feet higher than the other. The engineer changed the sidewalks in front of several stores to create a six inch step in the sidewalk, in addition to the curb. As a result of this change, a customer would walk out of a store on a level surface, and after walking about ten feet toward the street, the customer would encounter a sex step down the middle of the sidewalk. After walking another four feet, the customer would reach the street. Unfortunately, one customer, Donald Luther, fell on the step in the middle of the sidewalk and was hurt. He had climbed up the step while going into the store but says he forgot it was there when leaving. Apparently there was no handrail or marking on the step or sidewalk to indicate that there was a step. There was some indication that the step had been painted a bright yellow at one time but that the paint had worn off over the years.

In his case against the engineer, the plaintiff failed to present expert testimony concerning the standard of care that was owed by the engineer. Instead of expert testimony, the plaintiff presented testimony (mostly hearsay) that several other people had fallen in the same area. The court found that expert testimony was required to show what the standard of care was and that the engineer failed to meet the requisite standard. As explained by the court, expert testimony is required if the standard of care is not within the common knowledge of the jury. Only when a layperson would know based on their common knowledge that a professional service was negligent does it become unnecessary to have an expert. An example of such common knowledge of negligence would be where a surgeon cuts off the wrong leg of a patient. On the other hand, says the court, if there was a question whether the surgeon correctly performed a complicated surgery, an expert may be required.

Comment: This case once again demonstrates the importance in obtaining expert testimony to (1) establish the standard of care and (2) to prove that a design professional failed to meet that standard. In most claims against professional service providers for negligence, the types of questions for consideration by the jury go beyond the kind of common knowledge that a layperson would have as to whether the services were negligently performed. The number of cases like that of a surgeon cutting off a wrong leg being subject to a common knowledge determination of negligence are not nearly as common as the more subtle determinations that typically arise concerning whether a design professional exercised his or her services consistent with the generally accepted standard of care for similar services performed by similar professionals.

When filing a suit based on professional negligence, some states require an affidavit by an expert be attached to the complaint. Even where such an affidavit is not required, however, courts may dismiss a case for failure to present expert testimony. Moreover, even when a plaintiff believes the negligence is so obvious that it can be proved by lay testimony to the common knowledge of a lay person, it still may be most prudent to submit expert testimony rather than risk a court finding, as this court did, that the alleged negligence was not within a layperson's ability to determine without an expert.

Chapter 15

Federal Contracts

15.1 Hurricane Katrina's Impact on Existing U.S. Government Contracts (Vol. 7, No. 6)

By: *Dan Donohue and Hal Perloff*
Wickwire Gavin, P.C.

For contractors performing work on existing federal contracts in affected areas, assessing the impact of the hurricane on their contracts raises a number of issues and concerns. Federal construction and supply contracts typically provide for a non-compensable time extension for unusually severe weather or Acts of God. *See, e.g., The Rice Co.*, AGBCA No. 2003-188-1, 2005-2 BCA Sec. 32,005 (2005)(hurricane that delayed delivery of rice was excusable delay precluding the assessment of liquidated damages). However, under the "Permits and Responsibilities" provisions found in most federal contracts, a contract is responsible for repairing or rebuilding at its own cost any work damaged or destroyed by the storm that has not been accepted by the government.

In *DeRalco, Inc.*, ASBCA No. 41063, 91-1 BCA Sec. 23,576 (1990), a contractor was held responsible for the cost of rebuilding a 97.5% complete brick screen wall damaged by

Hurricane Hugo. The Board rejected the contractor's argument of defective government specifications (the wall was designed to withstand only 100 mph winds and not the 190 mph winds produced by Hugo) because the loss was caused by the hurricane and not the government's conduct.

A natural disaster such as Hurricane Katrina may also frustrate the purpose of the contract, making continued performance impossible or commercially impracticable. In these situations, the government may choose to terminate the contract for its convenience, entitling the contractor to be paid its costs to date plus certain costs of winding down the contract. Contractors should carefully follow instructions from the government and the procedures contained in the contract's termination for convenience clause. *See Dynatech Building Sys. Corp.,* ASBCA No. 47462, 95-1BCA Sec. 27,325 (1995) (a contractor forfeited its rights under the clause by filing its claim beyond the one year period provided for, even where its failure to submit its claim was caused by a hurricane).

Natural disasters may also cause the government to make changes in the work, entitling a contractor to equitable adjustments in the contract price and time. For example, a beach renourishment contractor was entitled to an equitable adjustment after a storm dramatically changed the contour of the existing beach and borrow areas and the contracting officer directed the contractor to change the locations where sand was to be deposited. *J.A. LaPorte,* IBCA No. 1014-12-73, 75-2 BCA Sec. 11,486 (1975). But the burden is on the contractor to prove the merits of such a claim, and relief will be denied where such proof is lacking. *See, e.g., L&C Europa Contracting Co, Inc.,* ASBCA No. 52848, 04-1 BCA Sec. 32,609 (2004) (under a contract to renovate a recreation center contractor failed to prove that it was damaged by delay to start of project due to roof damage caused by Hurricane Fran). Contractors are responsible for pursuing appropriate contractual relief for the effects of the disaster on their work.

The federal government may also assert warranty claims on existing buildings and other structures. Contractors should understand the extent of the warranties they have provided the government. It is the government's burden to prove its warranty claim. In many cases, there may be questions as to whether the specific weather conditions were within the scope of the coverage of the warranty. If government clients insist on purusing repairs under the warranty clause, contractors should be certain to receive a direction in writing from the contracting officer before starting work to preserve their rights to pursue compensation later, if justified.

Some contractor may be eligible to claim that they are entitled to extraordinary contractual relief under Public Law No. 85-804. Pursuant to that statute, executive agencies have the authority to enter into contracts and to modify existing contracts whenever that would facilitate the national defense. 50 U.S.C. Sec. 1431-1435, *see also* E.O. 10789, FAR Part 50.3. The granting of so-called "extraordinary contractual relief" is within the discretion of agency officials and is not a matter of right. FAR 50.301. The statute allows such relief when a contractor essential to national defense loses production capability. FAR 50.302-1(a). For instance, a contractor that was the only source of vital components to an ongoing military program might be able to gain 85-804 relief to repair its operations after Hurricane Katrina. There may also be arguments that the relief should be granted to contractors to provide indemnification for environmental liabilities on existing projects that have arisen as a result of the Hurricane.

About the Authors: Dan Donohue and Hal Perloff are attorneys with the law firm of Wickwire Gavin, with a law practice focusing on government contracts and construction law. They can be reached at ddonohue@wickire.com or hperloff@wickire.com, respectively, or 8100 Boone Blvd., Suite 700, Vienna, VA 22182; 703-790-8750.

Chapter 16

Fiduciary Duty

16.1 Project Manager Required by Fiduciary Duty to Owner to agree to Settlement with a Supplier Contrary to its Own Interest. (Vol. 7, No. 1)

Under the terms of a project management agreement making the project manager the owner's agent for dispute resolution of supplier claims was required to agree to a dispute settlement between the project owner and construction contractor that was adverse to the project manager's own interests.

Kvaerner U.S., Inc. ("Kavaerner") was responsible, pursuant to its project management agreement ("PMA") with IPSCO Steel, Inc. to recommend contracts for IPSCO to award to various subcontractors and suppliers, and to "be IPSCO's agent for the purpose of administering Supplier Contracts and managing and coordinating Suppliers' Work." In connection with liens and disputes, the PMA further provided that Kvaerner was "to protect IPSCO's interests at all times."

Under the PMA, Kavaerner expressly warranted that the "Aggregate Cost" of the project would not exceed a Guaranteed Maximum Price ("GMP") of $182 million and that it would reimburse IPSCO for any costs in excess of that amount. On the other hand, if the Aggregate Cost was below the GMP, Kavaerner was to share the savings on a 50-50 basis.

Another aspect of the PMA required Kavaerner to serve as IPSCO's litigation manager to resolve anticipated disputes that might arise with suppliers within the GMP. One final component of the PMA important to this decision was a requirement that IPSCO provide a $20 million professional liability insurance policy to cover Kvaerner, the subconsultants and design professionals. This policy was issued by Liberty Mutual Insurance Company and was a typical claims-made policy whereby defense costs erode the policy limit.

IPSCO awarded the design and construction contract to Blaine Construction Corporation ("Blaine") pursuant to Kvaerner's recommendation. Blaine abandoned the project, says the court, less than a year into its work due to design errors in its work which caused significant delays and disruptions. As a result of that abandonment of the work, IPSCO and Kvaerner entered into a written agreement reinforcing the agency relationship and specifying how any money recovered from Blaine would be paid to IPSCO and applied against the Aggregate Cost under the PMA. The court states that "IPSCO and Kavaerner agreed Kvaerner would pursue recovery from Blaine, Liberty Mutual and Marsh for damages resulting from Blaine's conduct."

In response to the suit brought against it by Kvaerner and IPSCO in the federal district court in Pennsylvania , Blaine demanded defense and coverage from Liberty Mutual. Liberty Mutual denied coverage on the basis that it allegedly had not received proper notice that Blaine was an insured under the policy. Blaine then filed suit against Liberty Mutual and Marsh, asserting that Marsh had issued an "advice of insurance" three years earlier assuring Blaine that it was covered by the Liberty Mutual policy.

A settlement was entered into between Blaine and IPSCO/Kavaerner whereby the parties agreed to submit the question of Blaine's liability to an arbitration panel and stipulated that if Blaine were found liable, a $26 million judgment would be entered against Blaine, but that because of Blaine's "empty pockets," IPSCO and Kavaerner would satisfy the judgment by looking solely to Blaine's insurers.

While all this was pending, IPSCO filed suit against Kvaerner in federal court in Alabama, demanding over $60 million in cost overruns, including damages from Blaine's abandonment of the project. Since Kvaerner was insured under the $20 million Liberty Mutual policy, its defense costs in the Alabama suit ($5million) were paid for by that policy, thereby reducing the coverage available to pay in the Pennsylvania litigation, if Blaine were eventually found liable and if Blaine prevailed in its coverage dispute with Liberty Mutual.

Liberty Mutual entered into a settlement agreement with IPSCO and Blaine while the arbitration was still pending. Kavaerner apparently declined to join the settlement discussions, says the court. By the terms of the settlement, IPSCO and Kvaerner would release all claims against Blaine and Blaine would release all coverage claims against Liberty Mutual, in payment of $6 million by Liberty Mutual. IPSCO then sent the settlement agreement to Kvaerner with the following transmittal language: "Pursuant to the PMA, IPSCO hereby directs Kvaerner, as its agent, to confirm in the space provided below that Kvaerner consents to the enclosed settlement agreement insofar as any such consent might be required from Kvaerner."

Kvaerner refused to sign the settlement agreement because it believed the $6 million settlement was not sufficient in view of the $26 million judgment that had been previously agreed to against Blaine. IPSCO then filed a motion with the Pennsylvania federal district court asking it to approve the settlement and dismiss all the claims, except those against Marsh which were resolved by a separate settlement agreement.

The district court concluded that Kvaerner couldn't unilaterally veto the settlement because, under the terms of the PMA, it was required to "protect IPSCO's interests" in any litigation with project suppliers. The district court held that Kvaerner had a fiduciary duty to act for IPSCO's benefit, that IPSCO had the right to control dispute resolution, and that Kvaerner was contractually obligated to follow IPSCO's instructions. And Kvaerner was barred by its PMA from putting its owner financial interests ahead of those of IPSCO.

On appeal, the U.S. Third Circuit Court of Appeals affirmed the district court decision. Of vital significance to the appellate court's decision was its finding that although the PMA gave "primary responsibility" for resolving disputes to Kvaerner, it reserves final settlement approval to IPSCO – thereby retaining the right to control settlement of disputes. The court found that "Kvaerner, as IPSCO's agent and pursuant to its agreement to protect IPSCO's interests was required to do IPSCO's bidding, which included Kvaerner's consenting to the two settlements.... As IPSCO's agent, Kvaerner owed IPSCO a duty of loyalty.... That duty of loyalty required Kvaerner to protect IPSCO's best interests. Once IPSCO made it known that it had reached a settlement with Liberty Mutual and Marsh, Kvaerner was under a duty to effectuate IPSCO's wishes and consent to settlements.... Kvaerner's duty of loyalty surmounted what could be considered as a conflict of interest."

The conflict of interest referred to by the court was explained as being the interest that Kvaerner had in ensuring that funds remained available under the Liberty Mutual policy to continue paying for Kvaerner's future defense costs in the Alabama litigation, in contrast with the interest of IPSCO to preserve as much of the policy as needed for the coverage action settlement in Pennsylvania. The court concluded: "Thus, it was in IPSCO's interests to reach a settlement with Liberty Mutual sooner rather than later. Even though such a settlement was adverse to Kvaerner's interests, Kvaerner was required, as protector of IPSCO's interests, to resolve any such conflicts in IPSCO's favor. *IPSCO Steel, Inc. v. Blaine Construction Corporation*, (No. 03-2929, U.S. Court of Appeals, 3rd Cir., June 10, 2004).

Comment: Several other interesting issues are discussed in this case, including whether the excess liability carrier, Lexington Insurance Company, had standing to intervene in litigation to assert that its own interests were adversely impacted.

Chapter 17

Indemnification

17.1 Indemnity Clause Requires Subcontractor to Indemnify Prime for Injuries Arising out of the Prime's own Negligence
17.2 Highway Contractor Protected by State Immunity Statute
17.3 Indemnification Clause Unenforceable if Negligent Parties Are Indemnified

17.1 Indemnity Clause Requires Subcontractor to Indemnify Prime for Injuries Arising out of the Prime's own Negligence (Vol. 7, No. 5)

Where the indemnity clause of a contract expressly exculpated a prime contractor from the consequences of its own negligence that resulted in injury to a subcontractor's worker, the prime was entitled to be indemnified by the subcontractor because the claim arose out of the performance of the contract.

In *Spawglass, Inc. v. E.T. Services, Inc.*, 143 S.W.3d 897 (Tex. 2004), the appellate court reversed a summary judgment that had been granted by the trial court in favor of the subcontractor. The contractor, SpawGlass Construction Corporation had subcontracted with E.T. Services,

Inc.("ETS") for ETS to perform structural steel erection for a high school. An employee of ETS, Brian Sanders, was working as a welder on the site. While he was rolling up an oxygen hose, he was struck by a sheet of plywood that blew off of the roof during a sudden storm. Sanders sued SpawGlass for negligence. SpawGlass in turn sought indemnity from ETS pursuant to the indemnity provisions of the contract.

SpawGlass contended that the contract clearly and unambiguously required ETS to indemnity SpawGlass for claims of injury to ETS's workers attributable to SpawGlass's negligence. ETS, in contrast, contended that the indemnity provision applied only to injuries resulting from ETS's performance. Flying plywood, says ETS, did not arise out of ETS performance. ETS argued that the indemnity may only be triggered if the incident arose out of its performance, not its mere presence on the site.

The appellate court rejected ETS's argument completely. First, the court found that the indemnity provision was clear and unambiguous with regard to meeting what is known as the "express negligence rule." That rule requires that the intent of the party seeking indemnity from the consequences of its own future negligence must be expressed in unambiguous terms within the four corners of the contract. In this case, the court held that the language clearly required that ETS would indemnify SpawGlass from the consequences of SpawGlass's own negligence that resulted in injury to ETS's worker.

With regard to whether the injury arose during ETS' "performance", the court held that ETS's argument that the injury arose from SpawGlass's *performance* completely unrelated to the work that ETS and its employee were hired to perform misses the point. The injury nevertheless occurred while all the parties were "engaged in the construction of a high school auditorium." Thus, the court concluded, "The claim asserted by Brian Sanders arises out of the *performance* of ETS's contract with SpawGlass. For these reasons, the appellate court reversed and remanded the trial court decision.

Comment: Based on the reasoning of this decision, it is important for parties that are negotiating indemnity provisions in contracts to carefully determine what they want to be indemnified, and to craft the language to accomplish that. As explained in this case, the "express negligence rule" that is applicable in most states means that if you want to be indemnified for your own negligence, you need to clearly state that intent in the contract. The contract in this case accomplished that for the prime contractor.

It is also not uncommon to see language like that in the contract at issue here which states that the indemnity applies to injuries or damages arising out of "performance of the contract." This does not necessarily mean that the injury has to arise directly out of the performance of the work performed by the party that is the Indemnitor. As explained in this case, just the fact that the worker was on the site because his employer was performing work for the prime contractor under a contract was enough to trigger the indemnity obligation. It didn't matter whether the employee or his employer had anything to do with causing the plywood to blow off the roof.

If you want to limit the indemnity to apply only to damages and injuries caused by your own performance, you can clearly state this in the contract. For example, if you are a design professional, you might state something to the effect that you will only indemnify the other party for damages "to the extent that they arise from the negligent acts, errors or omissions of the design professional." If you are a contractor, you might not be able to limit your indemnity to negligence based acts, but you might nevertheless limit your indemnity to apply only to damages "to the extent that they are caused" by you.

17.2 Highway Contractor Protected by State Immunity Statute (Vol. 5, No. 1)

Where a highway construction contractor followed specifications given to it by the state, it was immune from liability arising out of a motorist's personal injury action. The plaintiff's law suit alleged that her injuries were caused by, or made worse by, a guardrail that was constructed by the contractor. Under the state of Kansas law, a highway contractor is provided immunity from third party suits if certain conditions are met. These conditions are that (1) the injury occurred after the construction was complete; (2) the work of the contractor was accepted by the official responsible for the project; and (3) the contract provisions and specifications were satisfied by the contractor.

Plaintiff alleged that the contractor negligently performed its obligations under the contract by failing to install guardrails in accordance with the specifications, and that the contractor failed to warn KDAT that the guardrail it installed was negligently designed. Specifications for guardrails were provided by a KDOT document entitled "Protective Steel Plate Guard Fence at Bridge Piers." The guardrails depicted in the specification for single column and two-column piers was an "open-end" design, meaning that the ends of the guardrails on either side of the columns do not connect with one another. "Closed-end" guardrails were specified for certain other conditions as defined by the same document. The contractor installed the open-end design whereas the plaintiff asserts that a closed-end design was required by the specifications because the overpass contained a multiple-column pier. It is this discrepancy that forms the basis of the plaintiff's negligence claim and she asserts that it had been built closed-end, her car may not have flipped into the air when she hit the guardrail and she would not have sustained such serious injuries.

The contractor asserted that it fully complied with the specifications. The state supported the contractor's position by way of an affidavit filed by the assistant secretary of transportation. The contractor filed a motion for summary judgment to dismiss the complaint on the grounds that it was immune from suit pursuant to the statute. In the plaintiff's opposition to the motion, she included a counter-affidavit from an expert who opined that the closed-end design would have been required by the state at the accident site and that despite the state's acceptance of the contractor's work, the open-ended guardrails did not conform to the contract specifications. This affidavit was found by the district (trial) court to be too speculative and conclusory to establish a disputed issue of material fact. Accordingly, the court granted the contractor's motion for summary judgment.

On the appeal from that decision, the appellate court sustained the lower court decision in favor of the contractor and explained that the state immunity statute protected the contractor because the three-part test described above was met by the contractor. Specifically, the plaintiff's injuries occurred after the work had been completed; the state had accepted the contractor's work; and finally, no evidence was presented to establish a material factual dispute with respect to the contractor's compliance with the specifications. The court held that the purpose of the state statute was to protect contractors from liability when an accident occurs from completed highway project that the contractor had no role in designing. As seen by the court, "Such protection is logical, practical, and necessary." *Rodarte v. Kansas Dept. of Transportation, 39 P.3d 675 (Kansas, 2002).*

17.3 Indemnification Clause Unenforceable if Negligent Parties Are Indemnified (Vol. 6, No. 6)

Where an indemnification clause in a construction subcontract was so broad as to require the subcontractor to indemnify a project owner and construction manager for their own negligence, a court held the clause could not be enforced during a summary judgment motion requested by the indemnities. The clause could only be saved if it were proved that the indemnitees were not themselves negligent. That determination would have to await the outcome of the trial on the facts.

In *Lanarello v. City University of New York*, 774 N.Y.S. 2d 517 (2004), the court considered the enforceability of an indemnification clause that required the subcontractor to "indemnify the owner and construction manager [Morse Diesel] for any and all losses they sustain as a result of any or all injuries to any and all persons arising out of or occurring in connection with [subcontractor's] work, excepting only injuries that arise out of faulty designs or affirmative acts of the owner or construction manager committed with the intent to cause injury." The court concluded that this clause would indemnify the owner and construction manager for their own negligence and therefore "runs afoul of General Obligations Law section 5-322.1(1) of New York.

Morse Diesel, the construction manager ("CM") was asking the court to enforce the indemnity clause by way of a summary judgment motion to grant it judgment against the subcontractor. It argued that to the extent that the clause did not require the subcontractor to indemnify the CM for the CM's own negligence, the clause would be saved by another clause in the contract providing that "each and every provision of law and clause required by law to be inserted in the Contract shall be deemed to be inserted therein." In rejecting that argument, the court stated "Such language is not equivalent to language in the indemnification clause itself limiting a subcontractor's indemnification obligation 'to the extent permitted by law.'"

The Motion court that denied the summary judgment motion found that Morse Diesel had more than a mere general supervisory authority with regard to one of its subcontractor's who had responsibility for cleaning up debris and providing temporary protection around openings. Since negligence in those respects may have contributed to the accident, it would be necessary to allow the matter to go to trail so that it could be determined based on all the facts whether or Morse Diesel was negligent or not.

Comment: It is important to include a "survival" or "saving clause" directly inside the indemnification article so that if for any reason a court finds the indemnity language to be in violation of public policy or a state anti-indemnity statute, the article will nevertheless survive as language that falls back to that which is permissible under public policy and state law. This is often accomplished by introducing the article with language such as, "To the fullest extent permitted by law, the Contractor shall indemnify the Client ...." This may be more persuasive with a court than was the general saving clause that the court declined to apply in this particular case.

Chapter 18

License Requirements

18.1 Contractor Forfeited Right to Payment by Performing Work without a License (Vol. 7, No. 6)

Applying a contractor's state licensing law, a court held that where a contractor executed a contract before having its license and then obtained the license shortly after performance began, the contractor forfeited all right to payment for either the work performed before licensure or the work performed after licensure. Under a second contract with the same contractor for different work on the project, the contractor also had no license when it executed the contract but it obtained its license before beginning to perform any of its work under that contract. The court held that the contract was not null and void merely because the contractor did not have its contract as of the date of signing the contract, and that the contractor was therefore entitled to argue its right to payment for the work performed under that contract.

In *MW Erectors, Inc. v. Niederhauser Ornamental and Metal Works Company, Inc. (S123238)*, the Supreme Court of California, addressed the applicability of the Contractors' State License Law (CSLL; Bus. & Prof. Code, § 7000 et seq.). That statute imposes strict and harsh penalties for a contractor's failure to maintain proper licensure. Among other things, the CSLL states a general rule that, regardless of the merits of the claim, a contractor may not maintain any action, legal or equitable, to recover compensation for "the performance of any act or contract" unless he or she was duly licensed "*at all times* during the performance of that *act or contract*." (§ 7031, subd. (a) (section 7031(a)), italics added.)

The court explained that earlier case law softened the severity of this scheme by allowing contractors, though technically unlicensed at the time of performance, to show they had substantially complied with licensure requirements. But, says the court, the CSLL has since limited the availability of the substantial compliance exception by specifying that "[t]he judicial doctrine of substantial compliance shall not apply" unless the contractor "*had* been duly licensed as a contractor in this state *prior* to the performance of the *act or contract*" for which licensure was required.

The dispute in question arises out of hotel project being built for Disney Corporation by Turner Construction Company. Turner contracted with defendant Niederhauser Ornamental and Metal Works Company, Inc. (Niederhauser) to perform specialized metal work on the project, and Niederhauser, in turn, awarded two subcontracts to MW Erectors, Inc. (MW).

MW began work under the structural contract on or before the date it signed the contract. It did not obtain a C-51 structural steel contractor's license (see Cal. Code Regs., tit. 16, § 832.51) until about three weeks later. Under the separate ornamental contract, MW obtained a C-51 license before beginning performance of its work. An important side issue in the case was whether the fact MW didn't obtain a separate license specific to the category of "ornamental."

MW subsequently sued Niederhauser seeking alleged amounts due of $955,553 for work under the structural contract and $366,694 for work under the ornamental contract. Niederhauser moved for summary judgment, alleging that MW's claim was barred under section 7031(a), because MW had not been properly licensed at all times during the performance of its contracts. Niederhauser asserted that MW had no C-51 license when it began performance of the structural steel contract. Niederhauser also averred that MW could not demonstrate its substantial compliance with the C-51 license requirement because it had never held a California contractor's license before beginning work under the contracts in December 1999.

Niederhauser also argued that both contracts were illegal, void, and unenforceable *ab initio* because MW was unlicensed when they were executed. The Court of Appeal for the Fourth Appellate District, Division Two, held that the contracts were not void *ab initio* because of MW's unlicensed status when they were executed. Instead, said the appeal court, MW's right to recover depended on its licensure during its *performance* of the contracts. It held that MW could not recover for work it performed under the agreements during the relatively short time *before* it had secured a license, but that MW could obtain compensation for every individual act it performed under its contracts *after* all necessary licensure was in place.

Niederhauser sought review, urging that section 7031(a) required due licensure at all times during performance of a contract, and that both contracts were void *ab initio* because MW was not licensed when they were executed. The Supreme Court reviewed the statute at length, comparing it to an earlier version and also discussing numerous other court decisions that have applied the statute in different situations.

It is not possible in a short article like that which is offered in this Report to fully explain the Supreme Court's analysis and reasoning. But the essential conclusions of the court were these: (1) Where section 7031(a) of the statute applies, it bars a person from suing to recover compensation for *any* work he or she did under an agreement for services requiring a contractor's license unless proper licensure was in place *at all times* during such contractual performance; (2) Section 7031(a) does not allow a contractor who was at any time unlicensed during contractual performance to recover compensation for any individual *acts* performed while he or she *was* duly licensed; (3) a contractor who had not been duly licensed at some time *before beginning* performance under the contract may not assert protection under the substantial compliance exception to the strict enforcement of the statute; and (4) If a contractor is fully licensed at all times during contractual *performance*, the contractor is not barred from recovering compensation for the work solely because he or she was unlicensed when the contract was *executed*.

The Supreme Court concluded that "the Court of Appeal's interpretation contravenes well-entrenched case law. Prior decisions express a consistent understanding that one fails to meet the technical requirements now set forth in section 7031(a), and is ineligible to recover *any* compenstaiton under the terms of that statute, if, *at any time* during performance of an agreement for contractor services, he or she was not duly licensed."

The court went on to state as follows: "Addressing section 7031's plain language, we note first its specific provision that "no person . . . may bring or maintain any action, or recover in law or equity in any action . . . for the collection of compensation for the performance of any *act or contract* [requiring] a [contractor's] license" unless he or she alleges (§ 7031(a), italics added), and can prove (§ 7031, subd. (d)), his or her due licensure "at all times" during such performance (§ 7031(a)). The words "at all times" convey the Legislature's obvious intent to impose a stiff all-or-nothing penalty for unlicensed work by specifying that a contractor is barred from *all* recovery for such an "act or contract" if unlicensed *at any time* while performing it. This all-or-nothing philosophy is directly at odds with the premise that contractors with lapses in licensure may nonetheless recover partial compensation by narrowly segmenting the licensed and unlicensed portions of their performance."

The statutory language specifies that due licensure must have existed at some time "prior to" performance, and the court said that language cannot be squared with the notion that the contractor could first become licensed at some time *during* performance.

The court thus held that: "Because MW was not duly licensed "at all times" during performance of the structural contract (§ 7031(a)), and cannot alternatively establish its substantial compliance with the licensure requirements in that it had never held a valid California contractor's license "prior to" beginning performance… MW cannot sue to recover any compensation for work performed under that contract."

Insofar as related to this portion of MW's complaint, the summary judgment entered by the trial court was proper. On the separate issue of the validity of the ornamental metal work which was executed by MW before it had its license, the court affirmed the lower appeal court decision which had held that since MW had a contract prior to performing any of the work under that contract, it could present its case for compensation, because the contract was not rendered null and void solely because the contractor didn't have a license as of the date it signed the contract.

Chapter 19

Insurance

19.1 Waiver of Subrogation Enforced, Denying Insurance Company Recovery against Contractor it Alleged was Grossly Negligent
19.2 Faulty Workmanship Coverage Under CGL Policy
19.3 Pollution Exclusion in D&O Policy Applied to Exclude Coverage for Alleged Business Torts
19.4 Insurance Carrier not Required to Treat CM as Additional Insured Under Contractor's Policy
19.5 Broad Additional Insured Endorsement Entitles Contractor to Recover Damages under its Subcontractor's Primary and Umbrella Policies
19.6 Punitive Damage Award Against Insurance Company Reversed by Supreme Court as Excessive
19.7 Insurance Company that Incorrectly Denied Pollution Coverage Did not Act in Bad Faith in Failing to Defend and Indemnify its Insured
19.8 Insurance Coverage—Waivers of Subrogation

19.1 Waiver of Subrogation Enforced, Denying Insurance Company Recovery against Contractor it Alleged was Grossly Negligent
(Vol. 6, No. 7)

In an insurance case arising out a church fire that was caused by an employee of a contractor that had contracted to remove lead paint from and repaint the Church's exterior, the state supreme court held that Reliance National, and other insurance carriers of the church, were not entitled to subrogate against the contractor and a supplier. This is because a waiver of subrogation in the church's contract with the contractor was enforceable to bar the claim.

The contract between the church and contractor contained the following provision: ""The Owner and Contractor waive all rights against each other, separate contractors, and all other subcontractors for damages caused by fire or other perils to the extent covered by Builder's Risk or any other property insurance, except such rights as they may have to the proceeds of such insurance."

The fire was caused when an employee of the contractor, Knowles Industrial Services, Corp. (Knowles), brought a cigarette or open flame within ten feet of a section of the Church to which large quantities of the paint stripper had been applied earlier that day. The church was destroyed, with damages totaling almost $15 million. The church's insurance carriers paid it about half of those losses. Those carriers then sought to bring a subrogation suit in the church's name against Knowles and the manufacturers of the paint stripper that was used.

As to Knowles, Reliance's complaint alleged willful and wanton misconduct, negligence, breach of contract, and breach of warranty. As to the chemical defendants, the complaint alleged strict liability, negligence, and breach of warranty.

The lower appellate court granted a motion for summary judgment in favor of Knowles and the chemical defendants, and against the church/Reliance because the court found that the waiver of subrogation barred the claims

On appeal to the state supreme court, Reliance argued that genuine issues of material fact exist with respect to whether Knowles misrepresented its qualifications and intentions to comply with all pertinent federal and state regulations in order to obtain the contract from the Church. In reviewing this allegation the Supreme Court stated that as subrogee of the Church, Reliance is bound by the Church's statement of material facts and record references. Since the church did not argue or prove that Knowles made misrepresentations, the Court said this issue was not genuine and could not be presented by Reliance.

The court next dealt with the question of whether a wavier of subrogation is void as against public policy. As explained by the court, "A waiver of subrogation is a provision by which parties to a contract relieve each other of liability to the extent each is covered by insurance, thereby shifting the risk of loss to an insurer." The court further explained that it has previously held "waivers of subrogation are encouraged by the law and serve important social goals: encouraging parties to anticipate risks and to procure insurance covering those risks, thereby avoiding future litigation, and facilitating and preserving economic relations and activity."

In this case, Reliance argued that there must be a public policy exception to the general rule that waivers of subrogation are enforceable. Specifically, Reliance contended that public policy precluded the enforcement of the waiver of subrogation in this case based on Knowles's willful and wanton misconduct or its violation of a positive statutory duty, or because enforcement will be harmful to the interests of society.

The Supreme Court rejected all these bases. It acknowledged that exculpatory clauses are unenforceable in the face of claims of gross negligence or willful and wanton misconduct. But, the Court concluded "that principle is inapposite to waivers of subrogation." "The rule exists for exculpatory clauses to ensure that "a party injured by another's gross negligence will be able to recover its losses. [citations omitted]. In cases involving waivers of subrogation, however, there is no risk that an injured party will be left uncompensated, and it is irrelevant to the injured party whether it is compensated by the grossly negligent party or an insurer." The point that the court was making is that it makes a major distinction between a party indemnifying another for its own negligence and parties allocating risk to insurers.

What was particularly problematic to the Court with regard to Reliance's argument was that, "Adopting the approach advocated by Reliance would require us to distinguish between varying degrees of negligence. We have rejected the concept of gradations of negligence, [citation omitted] and we decline to change our approach with respect to waivers of subrogation for two reasons." The first reason given by the Court is that "waivers of subrogation deter litigation among parties to complicated construction contracts." In this case, the Court noted the real injured party in this case—the Church—was not a party to the appeal because the waiver of subrogation did what it was intended to do: it allowed the Church to resolve its claims quickly. The Church was made whole to the limits of its insurance and it was not divested of a remedy. "Were we to hold that parties cannot bar subrogated claims for gross negligence or willful and wanton misconduct, these benefits will evaporate, as the parties will have the incentive to litigate the question of whether a heightened standard of negligence applies."

The second reason given by the Court was that "waivers of subrogation have a beneficial economic effect that furthers the public interest. They help parties avoid the higher costs that result from having multiple insurance polices and overlapping coverage. [citation omitted]. In addition, because insurers can account for such waivers when setting premiums, [citation omitted], there is still an economic incentive for parties to refrain from committing gross negligence or willful and wanton misconduct." For these reasons, the Court concluded that "public policy favors enforcement of waivers of subrogation even in the face of claims of gross negligence or willful and wanton misconduct." *Reliance National Indemnity et al. v. Knowles Industrial Services, Corp. et al.* (868 A.2 220, 2005 ME 29).

Comment. This case highlights why project owners routinely seek waivers of subrogation as one of the terms of the construction contract. It also demonstrates why insurance carriers need to take these waivers seriously during the underwriting process—deciding whether to put limitations on them, or to charge additional premium for permitting the insured to grant the waiver of subrogation. Some carriers, particularly in professional liability policies, have granted greater leeway to their insureds to waive the right to subrogate. But the policy language states that such waivers may not be granted after the damage or claim, but may only be granted as part of the initial contract terms and conditions. Exercise caution before agreeing to any waiver. Be sure you know what your insurance policy permits. Discuss this with your insurance broker or carrier as appropriate.

19.2 Faulty Workmanship Coverage Under CGL Policy (Vol. 7, No. 4)

Costs of ripping out and replacing defective work was held to be potentially covered as property damage under a CGL policy issued by Zurich American Insurance, where employees of the insured contractor caused a leak in steam pipes by improperly unpacking the pipe prior to installation.

The contractor claimed that ripping out work of its various subcontractors, including the backfill subcontractor, the concrete subcontractor, and the landscape subcontractor, was all work performed on its behalf by subcontractors. This would make the rip out work covered, despite the defective workmanship exclusion that excludes coverage for property damage to "your work." The trial court ruled the damage was excluded but this was reversed on appeal.

Summary of the Facts: Limbach Company, LLC (Limbach) had a contract with Morse Diesel/Essex to perform mechanical work on a project at Howard University in Washington, D.C. Limbach was responsible for installing a prefabricated, insulated, underground steam line. It subcontracted the production of the steam line to Thermacor Process, Inc. (Thermacor). It subcontracted the excavation and backfilling of the trench for the line to Legacy Builders.

A leak was discovered in the steam line after it was installed. The leak damaged the insulation covering the pipe. It also damaged the backfill placed around the steam line and the landscaping in the area surrounding the leak – including concrete walkways. In order to excavate and repair the damaged pipe, Limbach had to remove concrete that was installed by a third party. As a result, Limbach had to hire a company to perform concrete replacement.

Limbach filed a claim with it commercial general liability (CGL) carrier, Zurich American Insurance, for the costs of replacing the damaged steam line and repairing the work damaged by the leak. This included the cost of repairing the backfill, the cost of replacing the steam pipe, the cost of repairing the landscaping, the cost of replacing the concrete, and the cost of a temporary steam boiler. Zurich agreed only to cover the cost of the temporary steam boiler and part of the cost of the landscaping. The balance of the claims were denied by Zurich on the basis of policy exclusions.

Relevant Insurance Policy Language: The Zurich policy provided that Zurich "will pay those sums that the insured becomes legally obligated to pay as damages because of 'bodily injury' or 'property damage' to which this insurance applies." The policy applied to completed operations, as defined by the policy under "products-completed operations hazard." This terms was defined in the policy as follows:

> " Products completed operations hazard: (a) Includes all 'bodily injury' and 'property damage' occurring away from premises you own or rent and arising out of 'your product' or 'your work' ...."

The definition of "your work" is "work or operations performed by you or on your behalf" and includes "materials, parts or equipment furnished in connection with such work or operations."

Exclusions to the policy included what is known as the "your work" exclusion. This excludes "'Property damage' to 'your work' arising out of it or any part of it and included in the 'products-completed operations hazard.'"

An exception to this exclusion provides: "This exclusion does not apply if the damaged work or the work out of which the damage arises was performed on your behalf by a subcontractor. Making this exception key to its argument for coverage, Limbach argued that its insurance claim covers the cost to repair or replace damaged work performed by subcontractors and third parties. Thus, Limbach argued that the damaged work was not excluded from coverage.

Choice of Law: As an initial matter, it is important to note that although the project was in Washington, D.C. and the court that decided the case was located in Virginia, the law the Court applied was the law of the State of

Pennsylvania because that is where the insurance policy was delivered.

Backfill Claim: The parties agreed that the backfill was damaged by the leak. The backfill work had been performed by a subcontractor. Zurich maintained that the "your work" exclusion precluded coverage for the damage to the backfill because it was performed on the insured's behalf. As work performed "on your behalf," Zurich argued this made it subject to the exclusion. The court rejected Zurich's argument because, "The exclusion specifically states that it 'does not apply if the *damaged work* or the work out of which the damage arises was performed on your behalf by a subcontractor.'" In reaching that conclusion, the court reviewed the history of the Insurance Services Office (ISO) exclusion on which this policy was based. The court quoted a decision from a Pennsylvania court that had analyzed the same exclusion and reached a "holding that the 'unambiguous terms' of the 'your work' exclusion do not eliminate coverage for harm done to a subcontractor's work."

In the present case, the court concluded that since the backfill was performed on Limbach's behalf by a subcontractor, the "your work" exclusion does not preclude coverage for the cost of repair to the damaged backfill. The Court said that "to hold otherwise would be to ignore the unambiguous terms of the exclusion's exception for work performed by a subcontractor.

The damaged pipe. With regard to the damaged pipe, Limbach argued that the damage was not excluded from coverage because the pipe was manufactured by a subcontractor, Thermacor. The lower court determined that Thermacor was a "materialman" rather than a subcontractor, and that the damaged steam pipe was therefore excluded from coverage by the "your work" exclusion. The appellate disagreed and found that Thermacor's role was highly distinguishable from that of a supplier because it had custom manufactured the steam pipe in accordance with shop drawings and project specification for this particular project. The court noted that one of Thermacor's representative visited the work site, reviewed the installation drawings with Limbach, and provided specific instruction regarding the installation of the pipe.

Replacing concrete and repairing damaged landscaping: The lower court applied the "your work' exclusion to preclude coverage for Limbach's costs of replacing the concrete and repairing the damaged landscaping that resulted from removing the damaged pipe. The appellate court reversed this and held that the exclusion does not exclude coverage for damage to a third party's work. "Since the landscaping and concrete work were performed by third parties, the "your work" exclusion does not preclude coverage for the costs of repairing and replacing the landscaping and concrete."

For these reasons, the appellate court reversed and remanded the district court's award of summary judgment that had been granted to Zurich . *Limbach Company, LLC v. Zurich North American* (CA-03-685-A, 4th Cir. U.S. Ct. App., Jan 2005).

Comment: This decision may have been decided differently by another court applying the law of a different state. When underwriting and pricing insurance policies, it is important for insurance carriers to consider differences in how courts in various states interpret the same language to reach very different results. It is important to consider where the insured will be performing its work and what state's law will apply. Many companies include a choice of law provision in the policy itself dictating that disputes between insureds and the insurance company will be decided in a particular jurisdiction which will apply the law of that jurisdiction. The decision in this case does not explain whether the policy included such a provision. One might conclude that when the court says the policy was "delivered" in Pennsylvania, it is also saying that it was issued pursuant to the law and regulations of Pennsylvania applicable to policies issued in that state.

Insured firms, such as construction companies, may also benefit from understanding differences in how the law of different states may affect the interpretation of both their insurance policy and their construction contract. It is interesting to compare court decisions that reach opposite conclusions concerning the intent of policy language. For a lengthy journal article analyzing in some detail issues and decisions surrounding coverage for construction defects, see John Lennes & Kent Holland, *Insurance for Construction Defects*, re-printed in Construction Risk Management Law and Case Notes, available at Amazone.com.

19.3 Pollution Exclusion in D&O Policy Applied to Exclude Coverage for Alleged Business Torts
(Vol. 7, No. 2)

When a property purchaser discovered pollution on its newly acquired property it made a claim against the seller, and the seller agreed to pay certain cleanup costs. But the seller subsequently reorganized its corporate structure and asserted the cleanup costs could not be paid out the reorganized company. The seller then brought suit alleging the seller intentionally and wrongfully reorganized its corporate structure to escape the liabilities. The seller asked its directors & officers insurance carrier to defend the suit. The carrier refused to do so because it claimed a pollution exclusion in the policy barred coverage. The seller filed suit against the carrier. After much litigation, the carrier's position was ultimately found correct by an appellate court which held that because there was a relationship between the purchaser's claims concerning wrongful reorganization and the pollution on the property, the policy's pollution exclusion must be applied to deny coverage.

The seller in this case was The Danis Companies ("Danis" or "TDC"). The purchaser was Waste Management, Inc. ("WM") . The insurance carrier was Great American Insurance Company ("Great American"). Great American refused to advance defense costs for Danis because it asserted that a pollution exclusion in the D&O policy barred coverage on the underlying claim against Danis by WM.

Waste Management had acquired the landfill from Danis as part of a sale of all outstanding shares of certain companies from Danis to Waste Management. In connection with the stock purchase, Danis agreed to indemnify WM against liabilities arising out of ownership of any landfill. The indemnity covered environmental liabilities, but it was not limited to just environmental liabilities.

As a result of pollution found after the transfer of the landfill, the Environmental Protection Agency ("EPA") issued notices of liability to Danis and WM as potentially responsible parties under Superfund. WM demanded indemnification from Danis. Danis eventually entered into a settlement agreement to indemnify WM for claims arising from environmental pollution, remediation, failure to remediate, toxic torts, bodily injury, and property damage. At some point, just before the settlement agreement was finalized, Danis underwent a major restructuring whereby Danis Building Construction Company ("DBBC") was separated from Danis Industries Corporation ("DIC") and The Danis Companies ("TDC").

Waste Management alleged in its litigation against Danis that DBBC had been a profitable subsidiary of TDC and was split off to insulate DBBC from environmental liabilities to WM. Waste Management also claims that the recapitalization/split-off of DBBC stripped DIC and TDC of assets, leaving insufficient funds to satisfy the indemnification obligations owed to WM under the settlement agreement. WM's complaint further alleged breach of several agreements by Danis concerning responsibility for the landfill remediation and liabilities.

Great American denied coverage for Danis and refused to advance costs for the WM lawsuit, asserting that the pollution exclusion of the policy barred coverage. The policy language stated that the policy excludes claims "based upon, arising out of, relating to, directly or indirectly resulting from or in consequence of, or in any way involving actual or alleged... pollution...; provided, however, that this exclusion shall not apply to any derivative suit by a security holder of the Company if the security holder bringing such Claim is acting totally independent of and without the solicitation, assistance, active participation or intervention of any Director or Officer of the Company."

Danis filed suit against Great American for breach of contract and declaratory judgment. The trial court agreed with Danis that coverage was not excluded by the pollution exclusion. The state appellate court reversed, however, because it concluded, "The use of the modifying words "directly or indirectly" indicates that an indirect causal relationship is sufficient for the exclusion to apply. Consequently, even though we have found that the federal claims are intertwined with the pollution settlements, coverage for these claims would additionally be excluded as matters 'indirectly related to ... pollution."

The reason the trail court concluded the pollution exclusion didn't apply was that it found the underlying federal claims revolved around allegations of corporate reorganization to escape liabilities. What the court essentially decided was that the alleged actions of the officers and directors in reorganizing and recapitalizing the corporate entities was the immediate cause of the harm alleged by

Waste Management, and that these actions served as an intervening cause between the pollution and the damages incurred by Waste Management. Thus, as an "intervening cause," the damages resulting from that cause would be independent of the pollution and would not be excluded by the pollution exclusion.

Great American argued that the trial court was wrong to apply an "intervening cause" approach. Although Great American admitted that the business torts create a separate cause of action, it argued that they are not independent causes of loss, because they could not have arisen in the absence of the underlying environmental liabilities.

The appellate court agreed with Great American that only one loss occurred—that being damage caused by the polluted site. "The acts of the Danis companies" concluded the court, "did not cause a separate injury or loss; instead, the alleged wrongful acts were an attempt to avoid paying for the loss. This is not the typical situation in which one party commits a tort, and the negligent or wrongful act of another party operates to cause either sharing or a complete release of liability for the injury." To be an intervening cause, explained the court, "the second negligent act must be both 'independent' and 'new....' The second act must not have occurred as a result of the first."

Applying these legal concepts, the appellate court found that the claims involved in the litigation by WM against Danis were not independent of the original pollution settlements but instead that the underlying settlements were part of the necessary predicate for liability of Danis is the federal case. The original pollution settlements and the alleged illegal transfers of assets are so intertwined and directly connected that the pollution exclusion is applicable to all the claims asserted. For these reasons, the court held in favor of Great American and reversed the trail court decision.

Danis v. Great American Insurance Co., 2004-Ohio-6222 (November 19, 2004).

19.4 Insurance Carrier not Required to Treat CM as Additional Insured Under Contractor's Policy
(Vol. 6, No. 6)

Where contractor was expected to make the construction manager ("CM") an additional insured under its general liability policy, but failed to do so, the contractor's insurance companies had no duty to provide coverage to the CM.

The contractor's insurance policies provided additional insured status to those with whom the contractor entered into written contracts. Where the CM, Morse Diesel, had no written contract with the contractor, it would not be an additional insured pursuant to the terms of the policies. In the case of *Lanarello v. City University of New York,* 774 N.Y.S. 2d 517 (2004), the court found that even if the CM were a third-party beneficiary of the contracts between the contractors and the project owner, that would merely give the CM standing to sue the contractors for breach of their contractual duty to the project owner to name the CM as an additional insured under their policies. This would not, however, create any independent duty of the insurance carriers to rewrite their policies to name the CM as an additional insured. Moreover, the fact that the insurance companies had issued certificates naming a predecessor construction manager as an additional insured, did not require them to treat Morse Diesel as an additional insured since that firm had not been specifically named as such and since Morse Diesel did not show that it had relied on the certificates that had been issued to its predecessor.

Comment: As indicated in the court's decision, a contractor's general liability insurance policy may state that parties with which it enters into signed contracts will be considered "additional insureds" under the policies. According to that language, third parties that are not in direct contract with a contractor will not be given additional insured status. For such a third party to become an additional insured, a request will need to be made of the insurance carrier and a certificate of insurance will need to be issued by the carrier to make a special exception to name such a third party as an additional insured. Unless that is done, the third party has no recourse directly against the contractor's insurance company. For these reasons, it is important to dot the I's and cross the t's when it comes to requesting and tracking certificates of insurance.

19.5 Broad Additional Insured Endorsement Entitles Contractor to Recover Damages under its Subcontractor's Primary and Umbrella Policies (Vol. 5, No. 9)

Contractor was entitled to recover as an "additional insured" under its subcontractor's primary and umbrella policies for damages suffered by a roofer who fell through a roof opening that had been cut by the subcontractor because the court found there was sufficient causal connection between the named insured's work and the situation that gave rise to the liability.

In this case of *Vitton Construction Co., Inc. v. Pacific Insurance, Co*, (California Superior Court, No. H205190-7, July 18, 2003), the general contractor, Vitton Construction Company (Vitton), entered into a subcontract with Pacific Erectors, Inc. (PEI), for "cutting and installation of roof opening frames." PEI was required by the subcontract to carry general liability insurance "covering all operations by or on behalf of [PEI] ... and including coverage for: (1) premises and operations; (2) products and completed operations; (3) contractual liability ...; (4) broad form property damage (including completed operations); (5) explosion, collapse and underground hazards; and (6) personal injury liability." PEI was further required by the contract to have its general liability policy name Vitton and the project owner as additional insureds.

One of the CNA Insurance Companies' divisions issued the primary policy to PEI , containing a "Blanket Additional Insured" endorsement. The endorsement defined "additional insured" as any person or organization PEI was contractually obligated to add as an additional insured, provided that such a party would only be considered an additional insured "with respect to liability arising out of ...' [y]our work' for that additional insured by or for you." Another endorsement of the primary policy named Vitton and the project owner as additional insureds with respect to "liability arising out of" PEI 's work on the project.

When an employee of a roofing subcontractor fell through one of the holes that had been cut by PEI, his claim for injuries was settled by the primary insurance carriers for both Vitton and PEI – both paying out their full policy limits. The balance of the settlement was paid by the AIU Insurance Company, Vitton's excess insurance carrier. The insurance company (Pacific Insurance) that provided an umbrella policy to PEI did not participate in the settlement and refused to contribute anything to the other carriers for their costs of the settlement.

AIU and Vitton then sued Pacific Insurance Company for subrogation and contribution on the ground that Vitton was an additional insured entitled to coverage under PEI's insurance policy with Pacific Insurance. Pacific argued in its defense that Vitton was not an additional insured because its liability did not "arise out of" work performed by PEI .

Key to the determination of whether Vitton was an additional insured under the Pacific policy was the Pacific policy language itself. It defined an insured as "any ... person or organization who is an insured under any policy of 'underlying insurance' ..., subject to all the limitations upon coverage and all other policy terms and conditions of such 'underlying insurance' and this policy."

Unless Vitton was an additional insured under the underlying policy, it would not by definition be an insured under the Pacific policy. The question in this instant case was whether the damages sustained arose out of work performed by PEI. If they did Vitton would be an additional insured.

In reviewing this question, the appellate court stated. "The California courts have consistently given a broad interpretation to the terms 'arising out of' or 'arising from' in various kinds of insurance provisions.... [I]t broadly links a factual situation with the event creating liability, and connotes only minimal causal connection or incidental relationship." Pacific argued that the facts of the case did not satisfy even this minimal level of causation.

In disagreeing with Pacific, the court stated the facts were not complicated and that it seemed fairly clear that the worker's fall arose out of PEI's work in cutting roof openings. It did not matter, said the court, whether or not it was PEI's responsibility to make the holes safe. The only relevant fact was that PEI's work created the condition that gave rise to the accident. It also did not matter whether or not PEI's work had been satisfactorily or negligently performed. "The fact that an accident is not attributable to the named insured's negligence is irrelevant when the additional insured endorsement does not purport to allocate or restrict coverage according to fault."

The court stated that if the insurance company had wanted to limit its coverage to damages that resulted from negligent performance of work it could have written its endorsement to so limit the coverage. Furthermore, the court concluded that "when an insurer chooses not to use such clearly limited language in an additional insured clause, but instead grants coverage for liability 'arising out of' the named insured's work, the additional insured is covered without regard to whether injury was caused by the named insured or the additional insured." Since the endorsement was not limited in this case, the general contractor, Vitton, was an additional insured under the various policies including the Pacific policy.

Comment: It is important to note the significant difference in the availability of additional insured status on a general liability policy versus a professional liability policy. Whereas additional insured status is routinely granted on general liability policies, it is seldom, if ever, granted under design professional liability policies. When, on those rare occasions, additional insured status is given to a project owner under the professional liability policy of its architect or engineer, the insurance company should certainly heed this court's observation about crafting specific language to clearly limit the coverage to damages caused by the negligent performance of the named insured of professional services on the particular project of the project owner who is the additional insured. The best solution, however, is for design professionals and their professional liability carriers to refuse to name project owners as additional insureds under professional liability policies. Project owners need to be consistently reminded and educated concerning the myriad reasons why it is inappropriate to name them as additional insureds.

19.6 Punitive Damage Award Against Insurance Company Reversed by Supreme Court as Excessive (Vol. 5, No. 7)

When State Farm Insurance refused a proposed settlement for the amount of its policy limit for a law suit arising out of an automobile accident, the matter went to trial and a jury returned a verdict against State Farm's insured in the amount of three times the policy limit. State Farm paid the entire judgment but the Insured (Mr. and Ms. Campbell) then sued the insurance company for bad faith, fraud, and intentional infliction of emotional distress. A jury in that trial awarded the Campbells compensatory damages of $2.6 million and punitive damages of $145 million.

The compensatory damages were eventually reduced by the state courts to $1 million but the $145 million punitive damages survived the appeals through the state courts. On appeal to the Supreme Court, State Farm argued that the punitive damage award was excessive and violated the Due Process Clause of the Fourteenth Amendment. The Supreme Court agreed and reversed the judgment, holding that under the circumstances of this case, it was more likely that the punitive damages that could be justified would be "at or near the compensatory damages amount."

The explanation by the Court is instructive. The court relied principally upon guide posts for reviewing punitive damages that it had established in the recent case of *BMW of North America, Inc. v. Gore, 517 U.S. 559.* "It should be presumed that a plaintiff has been made whole by compensatory damages, so punitive damages should be awarded only if the defendant's culpability is so reprehensible to warrant the imposition of further sanctions to achieve punishment or deterrence." The Supreme Court stated, "In this case, State Farm's handling of the claims against the Campbell's merits no praise, but a more modest punishment could have satisfied the State's legitimate objectives." Instead, the trial was used improperly as a platform to expose, and punish, the perceived deficiencies of State Farm's operations throughout the country. This resulted in the Utah courts awarding punitive damages to punish and deter conduct that bore no relation to the Campbell's harm.

The second guidepost considered by the court was the disparity between the actual harm suffered by the plaintiff and the punitive damages award. The Court stated that few awards exceeding a single-digit ratio between punitive and compensatory damages will satisfy due process. "Single-digit multipliers are more likely to comport with due process," said the Court. In this case, the Court found a presumption against an award with a 145 to 1 ratio. Applying The Gore case guideposts, the Court concluded that in light of the substantial compensatory damages award, it was likely that only punitive damages in about the same amount as the compensatory damages could be justified. *State Farm Mutual Automobile Insurance Co. v. Campbell, (No. 01-1289, April 7, 2003).*

Comment: The question of who gets to decide whether to settle a case has been a matter of concern for design professionals under their professional liability policies, and it has been a matter of concern for contractors under their various liability policies. Language from a typical professional liability policy states: "If YOU refuse to consent to any settlement or compromise recommended by US involving any part of OUR limits of liability and acceptable ot the claimant, and YOU elect to contest the CLAIM, suit or proceeding, then OUR liability shall not exceed the amount which WE would have paid for DAMAGES and CLAIM EXPENSES at the time the CLAIM or suit or proceeding could have been settled or compromised."

What is interesting is that professional liability carriers sometimes find themselves in the situation where they believe it is appropriate to settle a case but their insured design professional feels strongly that they did nothing wrong, and they don't want to settle for fear that this will be a black mark against them. This policy language is sometimes called a "hammer clause" and it says in essence to the insured design professional, "OK, you can exercise a right under the policy to refuse to settle the matter, but if the amount that is awarded at trial is greater than what we could have settled the case for, you (the design professional) will be responsible for the excess amount, and we (the insurance company) will pay only that part of the judgment that is within the amount for which the case could have been settled with the plaintiff.

The flip side of the settlement decision is what happened in the State Farm case. In the event that the carrier refuses to settle for an amount proposed by a plaintiff within the policy limit, as was apparently the situation in the underlying automobile claim that gave rise to the State Farm case, the carrier may be on the hook for the full judgment, including that part which exceeds the policy limit. But there is still the question of additional hardship, time, costs, and emotional distress incurred by the Insured as a result of the insurance company's refusal to settle the case when it could have been settled.

In an effort to encourage the carrier to settle (and for the purpose of laying a foundation for an eventual bad faith claim against the carrier), the insured's attorney will often write to the carrier at the time that the carrier is refusing to settle the underlying case. That letter will typically state that the Insured desires to settle the matter for the amount proposed by the plaintiff and that the insurance company's failure to do so is deemed by the insured to be bad faith and is subjecting the insured to damages such as those claimed by the plaintiff in the State Farm case. This may be considered by the court in determining whether the insurance company acted in bad faith, with intentional disregard of the best interests of its Insured under the policy.

19.7 Insurance Company that Incorrectly Denied Pollution Coverage Did not Act in Bad Faith in Failing to Defend and Indemnify its Insured
(Vol. 5, No. 3)

When an insurance carrier refused to defend the owners of a building under a general liability policy against claims by occupants alleging injuries from toxic fumes from carpeting, the owners sued the carrier to enforce their rights under the policy. They also sued for punitive damages, claiming that the insurance carrier denied coverage in bad faith. The courts held that despite an absolute pollution exclusion, the policy could not be applied to deny coverage for fumes from carpet glue. On the issue of bad faith, however, the court held that although the insurance carrier's failure to indemnify was wrong, it had a reasonable basis for incorrectly interpreting its policy and, therefore, it did not act in bad faith.

In *Freidline v. Shelby Insurance Company*, 774 N.E. 2d 37 (Indiana , 2002), the building occupants complained that substances that were used to install new carpeting in their offices caused them to become sick and to suffer bodily injuries. In response to a request by the building owners to defend them in the legal proceedings, and to indemnify them in case of judgment, the insurance carrier declined to do either. The owners then sued the carrier as stated above. On the issue of whether the pollution exclusion effectively precluded coverage, the appellate court concluded that the fumes emanating from carpet glue were not included in the policy's definition of pollutants, and that bodily injury arising from the fumes would therefore be covered.

The exclusion provided as follows: "This insurance does not apply to: ... Bodily injury and property damage arising out of the actual, alleged or threatened discharge, dispersal, seepage, migration, release or escape of pollutants... Pollutants means any solid, liquid, gaseous or thermal irritant or contaminant, including smoke, vapor, soot, acids, alkalis, chemicals and waste. Waste includes materials to be recycled, reconditioned or reclaimed." The court found this exclusion to be ambiguous and construed it against the insurance company so as not to exclude coverage for injuries arising from carpet glue fumes.

On the issue of bad faith, the court explained that in order to prove bad faith, the plaintiff would have to establish with clear and convincing evidence that the insurance carrier had knowledge that there was no legitimate basis for denying coverage. In seeking to prove this, the plaintiffs argued that the insurer knew of previous case precedent from the Court that had found the definition of pollutants to be ambiguous, and which had strictly construed the pollution exclusion against insurance companies.

The court acknowledges that in several reported decisions it did, in fact, find the pollution exclusion to be ambiguous, and that it had construed the language strictly against the insurance companies. The court noted, however, that the defendant insurance carrier in this case argued that each of those previous decisions dealt with business operations that involved the "handling and use of toxic or potentially polluting substances, so that the pollution exclusion would virtually negate coverage."

In contrast, the carrier argues that this case should be viewed differently because the insured building owner owns an office building which is an operation that does not regularly use toxic or caustic substances. They also argued that those previous cases involved environmental clean-up costs, whereas the instant case involves bodily injury to office workers. The insurance company cited to the court numerous recent out-of-state decisions holding that injuries resulting from similar types of emissions are excluded from insurance by the pollution exclusion.

Based on the good faith arguments and logic presented by the insurance company, the court found that there was a rational basis for the company's decision, even though that decision was wrong in the opinion of the court.. Consequently, the insurance carrier was entitled to summary judgment in its favor on the question of bad faith despite the court's conclusion that the company would be required to acknowledge coverage against damages arising out of bodily injuries.

19.8 Insurance Coverage—Waivers of Subrogation

(IRMI Expert Commentary, K. Holland - August 2002)

When an insured design-builder signs a contract with its client agreeing to a "waiver of subrogation, it gives up the right of its insurer to seek recovery for a third party (including the client) for the amount it paid to the policyholder for a loss caused by that third party. Many insurance policies state that the insured may waive subrogation provided it is done as part of the contract between the insured and its client and is done at the outset of the job and not after a claim or loss has arisen. The "waiver of subrogation" means that by giving up its right of recovery, the insurer accepts the fact that the policyholder and the parties with whom it has contracted have allocated the risk of the insured event to the insurer.

In *American Home Insurance Company, et al. v Monsanto Enviro-Chem Systems, Inc.*, 2001 U.S. App. LEXIS 17406 (4th Cir 2001), the Home Insurance Company filed suit against Monsanto Enviro-Chem Systems ("Enviro-Chem") to recover a loss that it paid under a property insurance policy on behalf of its insured, PCS Phosphate Company ("PCS"). Enviro-Chem had designed and built a chemical plant for PCS in 1985. In 1997 an implosion occurred at the plant, causing millions of dollars of damages to plant. American Home paid $5.6 million to PCS for the damage caused by the implosion and sought to recover that amount from Enviro-Chem.

American Home argued that Enviro-Chem knew of the risk that led to the accident and failed to warn PCS of the risk and remedial procedures that could have been implemented to avoid the risk. These allegations were based on the fact that a similar implosion had occurred in 1986 on another chemical plant designed and built by Enviro-Chem. Based on its investigation of the 1986 implosion, Enviro-Chem had recommended certain operational changes to the owner of that plant. In defending itself against American Home's suit, Enviro-Chem argued that the insurance company had waived its right of subrogation. The trial court judge agreed with Enviro-Chem and dismissed the action. This was appealed by American Home.

Fourth Circuit Court of Appeals

On appeal, the Court of Appeals for the Fourth Circuit concluded that American Home had no subrogation rights. The policy in question provided as follows:

> Owner [PCS] shall carry Builder's Risk Insurance "all risk" type coverage fully protecting Owner, Enviro-Chem ... against all physical loss or damage or damage to Plant, the Work, or any part thereof .... After such Builder's Risk Insurance shall have terminated, Owner shall maintain insurance covering, or assume the risk of, loss and damage to the Plant and the Work, *however caused*, and shall provide a waiver of subrogation in favor of Enviro-Chem ... under such insurance.

The lower court concluded that the term "however caused" was sufficiently broad and unambiguous to prevent the insurance company from having the right to assert claims based on negligent failure to warn. The appellate court rejected the insurance company's argument that the term "however caused" was ambiguous. Moreover, the court found that the subrogation clause explicitly applied to any damage to the plant or the work, which covered the entire heat recovery system.

Going a step further than the lower court holding, the appellate court analyzed the other terms and conditions of the contract between PCS and Enviro-Chem to determine whether American Home would have benefited by being able to "step into the shoes" of PCS if the subrogation right had not been waived. What the court found was that the design-build contract contained several clauses that would prevent PCS from recovering against Enviro-Chem for the damages in any event.

These clauses included a:

1. Waiver of consequential damages;
2. A limitation of liability; and
3. A clause stating that once Enviro-Chem completed its work under the agreement it would have no further obligation to PCS for any damage to the plant.

As a result of these terms and conditions, American Home could not have recovered against Enviro-Chem even if American Home had a right of subrogation since there was no contractual basis for a claim against Enviro-Chem.

Conclusion: This case demonstrates the importance of waivers of subrogation as part of the risk allocation strategy. The court considered the plain meaning of the "builder's risk" policy language, finding that the parties intended that insurance be the source of funds for the type of loss that actually occurred. The court also considered the plain meaning of the contract terms and conditions which specifically limited or barred causes of action by the owner against the design-build contractor—again indicating that the insurance would be the sole remedy and that the risk would be on the insurer.

Chapter 20

Insurance Coverage for Environmental Losses & Mold

20.1 Silica Claim Barred by Total Pollution Exclusion in CGL Policy
20.2 Broad Pollution Exclusion Is Ambiguous: Lead Covered by Policy
20.3 Whether Mold Cleanup Costs Are Covered Depends on Causation
20.4 Mold Loss Excluded under Homeowner's Policy – Summary Judgment for Carrier
20.5 Absolute Pollution Exclusion in Contractors Policy Does Not Bar Coverage for Injuries from Toxic Fumes

20.1 Silica Claim Barred by Total Pollution Exclusion in CGL Policy

(IRMI Expert Commentary, K. Holland - August 2005)

A commercial general liability (CGL) policy containing a "total pollution exclusion endorsement" was found to be effective in excluding claims based on alleged injuries arising out of inhaling silica dust from sand-blasting operations.

In *John Garamendi v. Golden Eagle Ins. Co.*, 127 Cal. App. 4th 480 (2005), the Court of Appeal of California affirmed a lower court holding that dismissed claims by plaintiffs who alleged they were exposed for many years to silica and silica dust at their employment, as a result of actions by 49 defendants. Among the defendants was Pauli Systems, Inc., who is alleged, collectively with the other defendants, to have:

> designed, tested, evaluated, manufactured, mined, packaged, furnished, supplied and/or sold abrasive blasting products, protective gear and equipment, safety equipment and/or sandblasting-related materials, equipment, products, etc.

Pauli Systems' CGL policy from the Golden Eagle Insurance Company had replaced a pollution exclusion that was standard in the policy as Exclusion f. with a total pollution exclusion that provided that the insurance would not apply to:

> "Bodily injury" or "property damage" which would not have occurred in whole or in part but for the actual, alleged or threatened discharge, dispersal, seepage, migration, release or escape of pollutants at any time.

The endorsement further defined pollutants as "any solid, liquid, gaseous, or thermal irritant or contaminant including smoke, vapor, soot, fumes, acid, alkalis, chemicals and waste...."

In response to plaintiffs' suit against it, Pauli Systems tendered the defense to Golden Eagle which denied coverage based on the pollution exclusion endorsement. Pauli Systems (hereinafter the "Claimant") then sued Golden Eagle, seeking a court order for coverage. Claimant argued that silica is not a pollutant because it is not smoke, vapor, soot, fumes, acid, alkalis, chemicals, or waste, and is found in commonplace materials such as sand, glass, and concrete.

In rejecting that argument, the court stated that even if silica is not one of the enumerated items of pollution in the policy, the listing is not exclusive. In addition, the court found that silica dust comes within the broad definition of "any solid, liquid, gaseous, or thermal irritant or contaminant." Moreover, pointed out the court, silica dust is identified by federal regulations to be an air contaminant. Thus, the court explained that:

> the widespread dissemination of silica dust as an incidental byproduct of industrial sandblasting operations most assuredly is what is "commonly thought of as pollution" and "environmental pollution."

Pauli Systems attempted to persuade the court that coverage might apply since part of the plaintiffs' complaint alleged a product defect. In comparing the language of the exclusion printed in the policy form and the exclusion of the endorsement, the court said the contrast in the language made clear "that under the operative endorsement in claimant's policy, there is no coverage for any of the claims of the underlying complaints, even if the products liability claims apply to claimant." The court held that:

> even on the assumption that claimant's alleged liability is based on the sale of defective products that contributed to personal injuries caused by silica dust, the injuries would not have occurred but for the discharge of the pollutant. Absent some other provision in the policy excepting product liability claims from the exclusion, the exclusion applies.

One final argument by the Claimant was that because the policy included an endorsement with a specific exclusion for claims based on exposure to asbestos, a reasonable insured party would understand that the pollution exclusion did not apply to claims for exposure to silica, for which there was no comparable explicit endorsement. The court rejected this argument with little discussion other than to say that in light of all the asbestos litigation that has been ongoing, "it is not surprising that an insurer seeking to exclude coverage for asbestos claims would include an explicit provision making that exclusion unmistakably clear." Significantly, however, the court concluded that:

> The inclusion of a specific provision concerning asbestos claims cannot reasonably be understood to mean that the pollution exclusion is inapplicable to other pollutants not specifically designated in a separate endorsement.

For these reasons, the court held in favor of the insurance company.

Comment: This decision is well-reasoned in its logic which grants to the insurance policy and its endorsements only the meaning reasonably intended by the insurer—the meaning easily and unambiguously understandable by a reasonable insured. It is most noteworthy that the court rejected Claimant's argument that by adding an endorsement explicitly excluding asbestos, the insurance company was required to add other endorsements to explicitly exclude items such as silica.

Insurance companies need the flexibility to sometimes reiterate, by issuing a separate endorsement, an exclusion for something they believe is already excluded under the general policy terms or under another exclusion. This is due, at least in part, because of surprising decisions by some courts that seem to stretch common sense as they attempt to find ambiguity in policy language and to find or even invent coverage for claimants where coverage was never intended by the insurer and never reasonably understood to exist by a reasonable insured at the time the policy was acquired.

Rather than asking courts to grant pollution coverage that was never intended, or paid for, in a CGL policy, a more appropriate insurance solution for a company with a known environmental risk may be to consider acquiring a separate policy specifically designed to cover its potential pollution liability.

20.2 Broad Pollution Exclusion Is Ambiguous: Lead Covered by Policy

(IRMI Expert Commentary, K. Holland - May 2005)

A federal court in New York found the language of the standard pollution exclusion of a commercial property insurance policy to be overly broad and ambiguous so as not to exclude coverage for lead dust resulting from a contractor's efforts to remove lead paint from a building.

In *Herald Square Loft Corp. v Merrimack Mutual Fire Ins.*, 344 F Supp 2d 915, (SD NY 2004), the question for the court was whether the pollution exclusion applied to lead poisoning that got into a building by lead paint dust blowing in through the windows when contractors sanded them. The property owner claimed more than $100,000 in cleanup expenses to clean up the lead paint dust contamination, replace window air-conditioning units and other equipment contaminated with the lead dust, and relocate some of the residents during the cleanup.

The court agreed that lead contaminants are a "pollutant" and that the lead contaminants were "released or dispersed" into the building. The court concluded, however, that the exclusion would not be applied in this case for the following reasons:

1. Overbroad language of the exclusion did not exclude coverage with the required specificity.
2. Applying the exclusion would not be consistent with "common sense and the reasonable expectations of the parties."
3. New York cases hold that lead paint is not an excluded contaminant.
4. The insurance company's notice of reduction in coverage adding a lead paint exclusion states an intent to reduce coverage and therefore suggests lead was covered prior to the addition of this new exclusion which was issued after the facts giving rise to the claim in this instance.
5. The insurance industry has left the relevant language in pollution exclusion clauses unchanged notwithstanding the numerous cases that question or reject its applicability to lead paint.

Exclusion Language

The policy in question contained an exclusion from coverage damage or loss arising out of the:

> discharge, dispersal, seepage, migration, release or escape of "pollutants" unless the discharge, seepage, migration, release or escape is itself caused by any of the "specified causes of loss." But if the discharge, dispersal, seepage migration, release or escape of "pollutants" results in a "specified cause of loss," we will pay for the loss or damage caused by that 'specified cause of loss.'

The term "pollutants" was defined in the policy to be:

> any solid, liquid, gaseous or thermal irritant or contaminant, including smoke, vapor, soot, fumes, acids, alkalis, chemicals and waste. Waste includes material to be recycled, re-conditioned or reclaimed.

The "specified causes of loss" for which pollution coverage was granted include:

> fire; lightning; explosion; windstorm or hail; smoke; aircraft or vehicles; riot or civil commotion; vandalism; leakage from fire extinguishing equipment; sinkhole collapse; volcanic action; falling objects; weight of snow, ice or sleet; and water damage.

Reasonable Expectations

According to the court: "The parties could not have reasonably expected the pollution exclusion ... to bar coverage for damages from repairs to the Building." The question, says the court, "is not whether leaded dust is a 'pollutant' for purposes of a pollution exclusion clause; it is whether a 'reasonable policyholder' would consider leaded dust removed from exterior windows and fire escapes during routine repairs to be environmental pollution." The court stated that the insurance company's interpretation of the exclusion clause is that the same language excludes coverage for damages from a large scale toxic environmental pollution as from repairs done to window sills. Without any explanation, the court concludes: "This could not have been the expectation of the parties when the ... policy was issued."

When the court says it "could not have been the expectation of the parties" to apply the exclusion equally to large scale pollution and dust arising from "repairs," it relies on previous New York cases finding the exclusion ambiguous with regard to coverage for dust and pollution from repair work. But the court fails to acknowledge that many insurance companies have intended and assumed that the exclusion would apply to all pollution regardless of size. What the court does, however, is state that because the insurance companies have left the relevant standard language of the pollution exclusion unchanged notwithstanding the numerous cases that question or request its applicability to lead paint, it must be concluded that "pollution exclusion clauses are inapplicable to losses resulting from lead paint unless such losses are specifically excluded."

The fact that the insurance company here eventually issued a "reduction in coverage" for a subsequent policy term seemed to the court to prove that the company considered damages from lead paint to be covered by the original policy language. Otherwise, reasons the court, why issue the exclusion? The court notes, however, that even if that was not the insurance company's intent, just the fact that the insurance company issued the separate lead paint exclusion prevents it from now arguing that the original exclusion was "clear and unequivocal" in excluding lead pollution.

Comment

One of the first things the court stated in its decision was this: "The language of the pollution exclusion clause of the 2002 policy is so broad that it cannot literally mean what it says." Really?! Does the court imply that the language should be taken figuratively? Are we to start looking for metaphors in the policy? Policy language is not the stuff of fiction and imagination. It is to be read and applied literally—accepting its plain meaning as its real meaning.

In view of the creative ways that insureds, their attorneys, and the courts have boldly found pollution coverage where insurance companies thought they had excluded it, perhaps a more foolproof way to successfully limit coverage would be to add a separate sublimit for pollution related to lead, asbestos, mold, and any number of other issues in which courts seem to favor finding coverage—even in the face of insurance company efforts to write pollution exclusions that are "absolute" or "total." If the words of the policy state that the most an insured will get for asbestos, lead, and mold is a specified, small dollar amount, it might be easier to make this stick in court.

The court used the fact that the insurance industry had not revised the standard form pollution exclusion following so many losses in the courts against the insurance carrier in this case. According to the court, this failure to revise the language must mean that the industry is satisfied with the coverage interpretations of the courts. How ironic, however, that any such revision to the standard form exclusions would logically be used by this same court against the insurers. Since this court found that adding a lead exclusion to the policy in question was evidence of the carrier's intent to cover lead under the original exclusion, wouldn't the court reach a similar conclusion if the standard exclusion itself was revised? By this court's reasoning, the insurers are damned if they do (revise the language), and damned if they don't.

There are times when an insurer may decide to revise its policy language—not because it wants to add a new exclusion, but because it decides to make the existing exclusion clearer. This might not be because the exclusion really needs clarification for the most people (i.e., the "reasonable man"), but because some courts have said it needs clarification. In light of court interpretations of exclusionary language that insurers thought was abundantly clear, insurers need the liberty to amend their forms, and even issue additional exclusions, without having courts then use this against them by either creating an adverse inference concerning the insurers' intent on the original language, or by concluding that the insurer cannot subsequently argue that the original language as "clear and unequivocal."

20.3 Whether Mold Cleanup Costs Are Covered Depends on Causation

(IRMI Expert Commentary, K. Holland - April 2005)

An insurer on a homeowners policy denied coverage for mold damages. It was held by an appellate court that if the homeowner proved the mold resulted from a covered peril, then the cost of removing the mold would be covered by the policy so long as it was not a loss separate from or caused by the covered peril or loss.

In *Simonetti v Selective Insurance*, 372 NJ Super 421, 859 A2d 694 (2004), a trial court found that there was no coverage under a homeowners policy for mold and other damages allegedly caused by water intrusion following a severe rainstorm in June 2001. The pertinent language of the policy contained an exclusion for "loss caused by ... mold" and for damages resulting from "faulty design ... workmanship ... [and] maintenance."

The Facts

It was about 2 months following the rainstorm that the homeowner discovered the mold and notified the insurer, Selective Insurance Company. The environmental claims unit of the insurer initially determined that the mold contamination and other water damages would be covered under the policy. Subsequently, however, the claims adjuster changed his mind upon learning that there had been water damage to the house several years earlier during 1997.

It appears that following the rainstorm damages, the homeowner initially made a claim against its homebuilder who tendered the claim to Royal Specialty Underwriting, its general liability insurer. Royal had a professional engineer inspect the house, and he wrote a report detailing his findings in September 2001. The report concluded that the water damage occurred from a time in 1997 when the homeowner had first noticed water intrusion around windows and had the builder caulk and otherwise repair the windows. The report further concluded that the leakage was due to a combination of design defects and waterproofing workmanship defects.

The homeowner had his own expert engineer also examine the house and write a report, dated October 1, 2001. That report also determined that the water intrusion resulted from poor workmanship during original construction of the house, and that, "The method of stucco and flashing application resulted in gaping holes in the wall permitting water entry into the wall cavity."

Based on the information from these two expert reports, the homeowners insurer in November 2001, issued a denial of coverage on the basis that the damage "resulted from wear and tear, deterioration, latent defect, inherent vice, corrosion, mold, wet or dry rot, settling including resultant cracking of walls; neglect, faulty, inadequate, or defective specifications, workmanship, construction, repair, materials used in repair, construction or maintenance." The homeowner responded by filing suit against the insurer for breach of contract and bad faith.

The Decision and Appeal

The trial court granted summary judgment for the insurer against the homeowner. This was reversed by the appellate court which concluded that mold damage caused by a covered event is covered under the Selective policy even though losses caused by mold may be excluded. The difference is whether the damages are caused directly from the covered event or from the mold itself. As stated by the court, "Mold can be both a loss and a cause of loss."

The court explained its reasoning as follows:

> This distinction between mold damage and loss caused by mold is supported by the very language of Selective's policy [which provides]: "we do not insure, however, for loss caused by ... mold...." This language clearly focuses on "cause" of the loss. But mold which is the loss is not mentioned. If Selective had intended to exclude not only losses caused by mold, but also mold itself, it could have easily expressed that intention.

Conclusion

As the court reverses and remands this case back to trial, it states that if the homeowner proves that the mold resulted from a covered peril, then the cost of removing the mold is not a different loss separate from or caused by the mold but rather is within the coverage for the basic loss. "In other words," says the court, "When a covered event causes mold, the mold damage includes the cost of removal."

It will be necessary for the trial (jury) to decide whether the mold and other damage claimed by the homeowner was caused by a covered peril or covered cause of loss. It is possible, explains the court, that two or more identifiable causes may contribute to a single property loss—with one of the causes being covered by the policy and one not being covered. This would not necessarily bar coverage since the policy did not contain an anti-concurrent or anti-sequential clause in its exclusions dealing with faulty design, workmanship, and maintenance. For these reasons, the appellate court remanded the case back to trial for a decision as to the actual cause of the damage (which will determine whether or not there is coverage).

20.4 Mold Loss Excluded under Homeowner's Policy Summary Judgment for Carrier (Vol. 6, No. 5)

A homeowner's suit against Lexington Insurance Company to recover under a property policy for damages caused by shoddy roofing work that resulted in mold contamination was dismissed based on a coverage exclusion for mold.

The insurance policy covered any risks of loss to the structure unless caused by one of the listed exclusions. It stated that "any ensuing loss to property described in Coverages A and B not excluded or excepted in this policy is covered." One of the listed exclusions was mold. The exclusion was specifically for loss caused by "smog, rust or other corrosion, fungus, mold, wet or dry rot."

The crux of the dispute concerned the question of whether the coverage for "any ensuing loss to property," in combination with the mold exclusion, should be read as excluding mold damages only when the losses were directly caused by mold. The homeowner argued that since the loss was caused by shoddy workmanship or vandalism by the contractor which would be a covered loss under the policy, the mold ensuing from that work or vandalism must also be covered. The court held that this position is not supported by the case law of the state.

In reviewing case precedents from several other states, the court explained that courts hold that "the ensuing loss provision does not reinsert coverage for excluded losses, but affirms coverage for secondary losses ultimately caused by excluded perils." In other words, if an uninsured peril causes a secondary loss of the type that is covered by the policy, that secondary loss will be covered only and this does not thereby result in the uninsured primary cause of that insured secondary loss being entitled to coverage under the policy as well.

In this case, the court held that mold, and the damage mold caused to the structure, are excluded. Because the court found that claims for damages resulting from mold damage to the structure were not covered, it granted summary judgment in favor of Lexington Insurance Company. *Brick v. Lexington Insurance Company*, (Superior Court of New Jersey, Docket No. ATL-L-1285-03 (April 2, 2004)).

Comment: This is an important decision in that it reiterates the intent of the insurance company that a mold exclusion is intended to exclude losses arising out of mold, regardless of how that mold was caused. Mold exclusions have routinely become a standard exclusion in most policies, either in the text of the base policy itself or in an endorsement added to the policy by the underwriter. The policy in question was a general liability property policy but the same concerns and exclusions are also seen in contractors' general liability policies and in the professional liability policies of design professionals.

A wide variety of mold endorsements have been drafted by the Insurance Services Organization (ISO) and numerous insurance companies. Some endorsements may provide mold coverage subject to a sub-limit. Others may provide a higher deductible for mold coverage than for the balance of the policy. There are "bodily injury only" mold endorsements as well as "property damage only" endorsements. And there are endorsements granting mold coverage provided it does not result from improper maintenance or in other cases from faulty workmanship or defective design. Some endorsements may grant mold coverage on certain types of commercial facilities but exclude it on residential facilities. The possibilities are extensive. This is definitely not a situation where one size fits all. By working with its insurance broker and insurance company, the insured may be able to obtain coverage to at least cover some of its risk arising out of mold.

20.5 Absolute Pollution Exclusion in Contractors Policy Does Not Bar Coverage for Injuries from Toxic Fumes

(IRMI Expert Commentary, K. Holland – Nov. 2005)

In a case addressing the applicability of a pollution exclusion provision in a commercial general liability (CGL) insurance policy, the New Jersey Supreme Court held that the absolute pollution exclusion applied only to traditional pollution claims and could not be used by the insurance company as a bar to coverage for personal injury allegedly caused by the exposure to toxic fumes that emanated from a floor coating/sealant operation performed by the insured contractor.

The insured contractor was NAV-ITS, Inc. (Nav-Its). It specialized in tenant "fit-out" work, including the building of partitions, the laying of concrete, the installation of doors, and the application of finishes, such as paint, sealants, and coatings. Nav-Its entered into a contract to perform fit-out work at a shopping center and had a CGL from defendant Selective Insurance Company of America (Selective) for its work. Nav-Its hired T.A. Fanikos Painting (Fanikos) as a subcontractor on the project to perform painting, coating and floor sealing work. While the work was underway, a doctor (Roy Scalia) with office space in the shopping center was allegedly exposed to fumes that were released while Fanikos performed the coating/sealant work. As a result of that alleged exposure, Dr. Scalia claimed that he suffered from nausea, vomiting, lightheadedness, loss of equilibrium, and headaches.

The doctor filed a complaint against Nav-Its and several others for personal injuries arising out of his exposure to fumes at his office. Nav-Its forwarded the complaint to Selective, seeking defense and indemnification. Selective refused to provide coverage to Nav-Its. In binding arbitration, Nav-Its was found liable to the doctor.

Following its loss to the doctor, Nav-Its sued Selective in a declaratory judgment action, asserting that Selective breached its duty to defend and indemnify it against the doctors claim.

Selective's insurance policy contained a pollution exclusion endorsement that provided in relevant part: "[Selective] shall have no obligation under this coverage part:

a. to investigate, settle or defend any claim or suit against any insured alleging actual or threatened injury or damage of any nature or kind of persons or property which:

1. arises out of the 'pollution hazard:' or

2. would not have occurred but for the 'pollution hazard:' or

b. to pay any damages, judgments, settlements, losses, costs or expenses of any kind or nature that may be awarded or incurred by reason of any such claim or suit or any such actual or threatened injury or damage; or

c. for any losses, costs or expenses arising out of any obligation, order, direction or request of or upon any insured or others, including but not limited to any governmental obligation, order, direction or request, to test for, monitor, clean up, remove, contain, treat, detoxify, neutralize, in any way respond to, or assess the effects of 'pollutants.'"

The policy defined pollutants as "any solid, liquid, gaseous, or thermal irritant or contaminant, including smoke, vapor, soot, fumes, acids, alkalis, chemicals and waste." Under the policy, "[w]aste includes materials to be recycled, reconditioned or reclaimed." It also defined "Pollution Hazard" to mean "an actual exposure or threat of exposure to the corrosive, toxic or other harmful properties of any 'pollutants' arising out of the discharge, dispersal, seepage, migration, release or escape of such 'pollutants.'"

A limited exception to the pollution exclusion was provided by the policy so that the pollution exclusion would not apply to:

"B. Injury or damage arising from the actual discharge or release of any "pollutants" that takes place entirely inside a building or structure if:

1. the injury or damage is the result of an exposure which takes place entirely within a building or structure; and

2. the injury or damage results from an actual discharge or release beginning and ending within a single forty-eight (48) hour period; and

3. the exposure occurs within the same forty-eight (48) hour period referred to in 2. above; and

4. within thirty (30) days of the actual discharge or release:

a. the company or its agent is notified of the injury or damage in writing; or

b. in the case of 'bodily injury,' the 'bodily injury' is treated by a physician, or death results, and within ten

(10) additional days, written notice of such injury or death is received by the company or its agents.

Strict compliance with the time periods stated above is required for coverage to be provided."

Nav-Its argued that the applicability of the pollution exclusion clause should be limited to traditional environmental type claims because this is consistent with the majority of decisions from other jurisdictions. Nav-Its also argued that its position is consistent with its reasonable expectations that coverage would be provided for claims arising from the normal work of the insured for which it was seeking insurance coverage, and that the purpose of the pollution exclusion was to preclude coverage for claims involving environmental contamination.

Selective argued in response that the pollution exclusion, by its plain terms, is not limited to traditional environmental claims. It claimed that insurance companies and policyholders are entitled to have their rights protected by the courts, and when insurance companies draft clear, unambiguous policies, the policy should be enforced as written.

In analyzing the policy language, the court explained that if policy language is clear, the courts will give the policy's words their plain and ordinary meaning. But where the language is ambiguous, the policy will be construed in favor of the insured and against the insurance company. The court stated that: "Because of the complex terminology used in the policy and because the policy is in most cases prepared by the insurance company experts, we recognize that an insurance policy is a "contract[] of adhesion between parties who are not equally situated."

The court further stated, "... courts also [] endeavor [] to interpret insurance contracts to accord with the objectively reasonable expectations of the insured."

The court proceeded in its analysis to do a lengthy review of the development of the pollution exclusion. From that review, the court concluded that the exclusion was only intended to exclude what it called traditional environmentally related damages. The court also reviewed the history of how the exclusion was presented and explained to the state insurance commissioner for approval. It concluded as follows:

"[T]here is no compelling evidence that the pollution exclusion clause in the present case, when approved by the Department of Insurance, was intended to be read as broadly as Selective urges. See Stempel, supra, 34 Tort & Ins. L.J. at 33. ('If the absolute exclusion was intended to reach as broadly as now contended, one would expect to see conclusive ISO memoranda and similar documents'). To be sure, read literally, the exclusion would require its application to all instances of injury or damage to persons or property caused by 'any pollutants arising out of the discharge, dispersal, seepage, migration, release or escape of . . . any solid, liquid, gaseous, or thermal irritant or contaminant, including smoke, vapor, soot, fumes, acids, alkalis, chemicals and waste.' If we were to accept Selective's interpretation of its pollution exclusion, we would exclude essentially all pollution hazards except those falling within the limited 'exception' for exposure within a structure resulting from a release of pollutants 'within a single forty-eight hour period.' We reject Selective's interpretation as overly broad, unfair, and contrary to the objectively reasonable expectations of the New Jersey and other state regulatory authorities that were presented with an opportunity to disapprove the clause."

The court was critical of the insurance industry for presenting the absolute pollution exclusion to the Insurance

Department as mere clarification of the policy intent when in fact the insurance industry is now asserting what the court believes to be a major change in the coverage. Moreover, the court suggested that with such a change, the carriers would have been required to reduce the premium rates or seek approval from the Insurance Department for its rates.

The court concluded: "Rather than 'clarify' the scope of coverage, the clause virtually eliminated pollution-caused property-damage coverage, without any suggestion by the industry that the change in coverage was so sweeping or that rates should be reduced. For those reasons, we decline to enforce the [] pollution-exclusion clause as written. To do so would contravene this State's public policy requiring regulatory approval of standard industry-wide policy forms to assure fairness in rates and in policy content, and would condone the industry's misrepresentation to regulators in New Jersey and other states concerning the effect of the clause."

SDSFor these reasons, the court held: "Because we conclude that the pollution exclusion clause as presently approved should be limited to traditional environmental pollution, we disapprove of any contrary view expressed in our case law." *NAV-ITS, Inc., v. Selective Insurance Company of America* (NO. A-20/21. N.J. Supreme Ct., 2005)

Comment

In reaching its unanimous holding in this case, the New Jersey Supreme Court stated that its decision to limit the pollution exclusion to those hazards traditionally associated with environmentally related claims is consistent with the highest courts in California, Illinois, Massachusetts, Ohio, New York and Washington. There certainly appears to be a growing trend for courts to interpret the pollution exclusions against insurance carriers to limit their effectiveness and impact. As an alternative to relying solely upon exclusions, another possibility might be to craft policy language to establish a separate sub-limit for any all claims arising out of pollutants (with pollutants defined even more broadly). This could also be done for claims arising out of or related to mold. At least in this way, the insured would be harder pressed to assert that it didn't know it had no coverage or only very limited coverage – especially since the sub-limit would be stated clearly on the Declaration Page of the Policy – easily seen and understood by the insured and its broker.

Insurance companies can be caught between the proverbial "rock and a hard place" when courts, like this one, criticize them for the way they present and describe to state insurance departments their new exclusions (such as ISO language exclusions). The exclusion in question was not intended to "change" the insurance carrier's intent from the previous language. The exclusion was, as the carrier here claimed, intended to merely clarify what had already been excluded in the previous language. It was to restate what carriers had intended their language to mean all along so that even the most creative plaintiff's attorney could not confuse courts on the issues.

The problem for the carriers that led to the new language was that earlier courts interpreted the language contrary to their intent. In response to those decisions the carriers felt erroneous, care had to be taken in how to proceed with new policies. If the carrier did nothing to revise its language in response, the court might conclude the carrier was acquiescing to the court's interpretation. See, for example, the case cited in my March 2005 IRMI Expert Commentary article discussing the New York case of *Herald Square Loft Corp. v Merrimack Mutual Fire Ins.*

So what is a carrier to do? If the carrier were to go the insurance department and say, "We're changing the language to reduce coverage from what we previously intended," the plaintiffs' attorneys and courts would jump all over that as a basis to grant coverage for all kinds of plaintiffs that had policies with the previous language. This would be called an "admission against interest." It would be used as evidence in court against the carrier in litigation involving claims arising under old policies. So, it's pretty obvious the carriers can't do that. But according to the New Jersey court, the carrier also cannot go to the insurance department and say, "We're changing the language merely to try to make it perfectly clear (even to plaintiffs attorneys and judges who love plaintiffs and dislike insurance carriers), that we are not covering these types of alleged losses under our policy nor have we intended to cover them under the previous language." Yet, that seems like precisely what is needed, and that courts should not penalize insurance carriers by finding mere clarifications to be reductions in coverage.

Chapter 21

Limitation of Liability

21.1 Limitation of Liability Clause Protecting Owner is Not Voided by Owner's Breach of Contract or Alleged Bad Faith
21.2 Liquidated Damages Clause and Waiver of Consequential Damages Clause Effectively Cap Damages Available against Design-Builder

21.1 Limitation of Liability Clause Protecting Owner is Not Voided by Owner's Breach of Contract or Alleged Bad Faith (Vol. 5, No. 4)

A Limitation of Liability (LoL) clause in a contract was upheld by a court notwithstanding allegations that the project owner had acted in bad faith in its treatment of the contractor. It was held to apply, however, only to the damages that would be awarded under the contract and not to limit additional damages for interest, attorneys fees, and other costs that were imposed under state statute.

Where a painting contractor and the project owner, Sun Company, could not agree on inspection standards and whether the contractor's paint stripping met the contract specifications, the contractor left the job and Sun eventually issued a letter to cancel the contract. Sun offset its costs of re-procurement and completion of the paint job against the balance claimed by the contractor for work it had performed. The contractor filed suit to recover the balance of what it thought was due for the work that had been performed. The trial court trial court rejected Sun's claim that any remedies were subject to the contract's LoL clause because it found Sun had acted in bad faith.

In reviewing the matter, the appellate court stated that LoL provisions are not disfavored by the state and that such clauses are binding on parties unless they are unconscionable. Regardless of whether there was an unjustified breach of contract, the court explained that by their contract language parties may agree to waive remedies that they would otherwise have under contract law. The court's decision suggests that this could be applied to both statutory and common law remedies if the LoL clause was clearly drafted to express that intent.

In determining the impact of Sun's breach of its implied duty of good faith inspection on the contract's other provisions (such as the LoL clause) the court reviewed the contract as a whole. It found significant the fact that the contract contained multiple provisions permitting Sun to "terminate," "cancel," or "suspend" the contract at its sole discretion for any reason -- or for no reason whatsoever. The appellate court concluded that Sun had the right to terminate the contract and was not required, as the trial court had wrongly concluded, to try to work out with the contractor its dispute over the inspection and the quality of the work being performed. Nevertheless, the court found that the trial court's error was harmless in that the contractor was indeed entitled to recover its costs and fees under the contract – even as terminated, and that the award of the trial court had been within the amounts permitted under the LoL clause which limited contractor recovery to the total contract price. *John B. Conomos, Inc. v. Sun Company, Inc.*, 831 A.2d. 696 (Pa. 2003).

Comment: The court's discussion of the interpretation and enforceability of the LOL clause in the contract demonstrates several points for consideration when drafting LOL clauses. These clauses are often enforced even in the face of difficult facts or allegations when both parties are commercial enterprises as was the situation here. The clauses can limit recovery that would otherwise be permitted under the law of the state but to do so, they must clearly express that intent. In this case, the clause did not expressly state that interest and attorneys fees would be affected by the clause and the court declined to apply it to these remedies that were imposed by statute rather than by the contract.

As a general matter, it may be prudent to keep the LoL clause separate from an Indemnification clause. Whereas state anti-indemnity statutes may restrict the use of an indemnification clause, the same statute might not restrict the use of an LoL clause. A court that may be inclined to find an indemnification clause to violate public policy may be less likely to find fault with an LoL clause that parties bargained for and that only affects their rights as against one another.

21.2 Liquidated Damages Clause and Waiver of Consequential Damages Clause Effectively Cap Damages Available against Design-Builder
(Vol. 5, No. 2)

Contracts requiring a design-build engineering firm to supply "basic engineering packages" for licensing and technology transfer agreements for the design and construction of a processing plant for sodium hydroxide (caustic soda) contained a liquidated damages clause capping the engineer's liability at 10 percent of its fee, and also contained a waiver of consequential damages clause waiving "special, indirect, incidental, or consequential damages of any kind." In response to the project owner's suit against the engineer for failure of the plant to achieve commercial production, the court enforced these clauses to limit the available recovery.

The plaintiff's complaint against the contractor alleged breach of contract, misrepresentation and fraud. With regard to the counts of the complaint alleging misrepresentation and fraud, the court dismissed these because they were barred by the two year statute of limitations. In response to the defendant's argument that the breach of contract claim should also be dismissed based upon the Waiver of Consequential Damages and the Liquidated Damages clauses, the plaintiff argued that the clauses should not be enforced because the clauses were unconscionable, were based on material misrepresentations, and were the product of mutual mistake.

The Waiver clause provided: "Article XV Waiver of Consequential Damages. In no event shall Seller [contractor] be liable to [owner] whether in contract, warranty, tort (including negligence or strict liability) or otherwise for any special, indirect, incidental or consequential damages of any kind or nature whatsover."

The liquidated damages clause provided: "Article VIII Liquidated Damages. In the event that the Caustic Prill Unit fails to produce Caustic Soda beads during the performance test even though all the conditions described in Article VII hereof have been satisfied and despite [contractor's] efforts to correct said failure, for each 5 percent or part thereof shortfall below the level warranted in Article VII, hereof, [contractor] will pay to [owner] an amount equal to 5 percent of the lump sum fee received by [contractor] for the failed Caustic Prill Unit. However, [contractor's] maximum limit of liability under the Agreement as to any failed Caustic Prill Unit shall be 10 percent of the Lump sum fee received by [contractor] for the failed Caustic Prill Unit. These payments are the exclusive remedies provided to [owner] under this Agreement. Except as provided in the Article VII, Contractor shall have no other liability whether in contract, warranty, tort, or otherwise."

The plaintiff, project owner, tried to get around the liquidated damages clause by arguing that it only applied in the event that the Unit failed the performance test. Since there was never a performance test, it argued the limitation clause had no effect. In interpreting the contract on this matter, the court explained that "the intention of the parties is a paramount consideration." Intent must be ascertained from the contract document itself when the terms are clear and unambiguous.

The court concluded that the clause makes clear that although the five percent cap appears to apply in the event of a performance test failure, the ten percent cap applies to any claim under the Agreement regardless of whether or not performance tests were performed. The court emphasized that "When combined with the extremely strong liability-limiting language of the entire clause, these phrases make clear that the intention of the parties was to limit [owner's] recover under any circumstance to ten percent of the fee it paid to [contractor]."

The court also rejected the project owner's argument that the clauses were "unconscionable" and should not be enforced. The court said that the test under Pennsylvania jurisprudence for unconscionability is "an absence of meaningful choice on the part of one of the parties together with contract terms which are unreasonably favorable to the other party." It further explained that the principle underlying the concept is to prevent oppression and unfair surprise but that it is not intended to disturb the "allocation of risks because of superior bargaining power." In other words, just because a party has greater bargaining power and negotiates a more favorable and even onerous deal does not make the deal unconscionable in the absence of oppression and unfair surprise. In commercial settings, explains the court, a limitation of damages clause will rarely be found unconscionable.

In this case, the owner claimed that it was a small unsophisticated Indian company that trusted "an American behemoth" when its president flew to Philadelphia to sign the deal. It made no changes to the contract and did not seek counsel to assist with its negotiation. Although the court described this as a "sympathetic picture," the court concluded that the scenario did not suggest any lack of meaningful choice.

In its conclusion with regard to this issue, the court said, "There is nothing in the record to suggest unfair surprise.... The clauses were not hidden boilerplate. The one point which gives this Court pause is whether a ten percent cap creates an adequate incentive to perform. However, there is no indication that the profit margin was any higher than ten percent. Therefore, [owner] has not demonstrated unconscionability." *Mistry Prabhuda Manji Eng. Pvt. Ltd. v. Raytheon Engineers & Constructors, Inc., 213 F.Supp.2d 20 (U.S D.C., Massachusetts, 2002).*

Comment: This case provides valuable insight into the judicial interpretation and application of contract clauses that purport to limit liability of engineers and contractors. There is a striking similarity in the project owner's arguments with those that have been raised in so many other reported cases. This decision should be a reminder to every commercial entity entering a contract for the design or construction of a project that, generally speaking, courts will enforce the terms of the contract that result from arms-length negotiations between two commercial entities. This is true even if one of the parties was significantly smaller than the other and did not have equal bargaining clout.

The key, as explained by this court, is whether the damage limitations would be unconscionable. In my own legal practice, I have had more than one client tell me that they wanted to ignore my advice and sign onerous contracts in which they would to be giving away substantial rights to the other party – with the expectation that they could convince a court that they signed the contract as a result of duress, coercion or unequal bargaining position and that the clause should be void as against public policy or as unconscionable. My advice has been that a court would not be impressed with their arguments for much the same reasons stated by the court in this case. Plus, my clients have had competent legal assistance with their contracts and this makes their chances of getting a court to let them out of a bad deal even more unlikely.

Note, however, that the court provides significant pointers in drafting an enforceable limitation of liability clause, when it states that the clause in this case was not "hidden boilerplate" and that the question of whether a ten percent cap creates an adequate incentive to perform gave the court pause. I typically advise clients to make clauses such as indemnification, limitation of liability (LoL), and waiver of consequential damages clear and pronounced in the contract. If an LoL clause might be subjected to close judicial scrutiny it may even be advisable to have your client separately initial or sign their name beside the clause so they cannot later claim they were surprised to learn of its presence in the contract. In addition, you should be careful to make the LoL amount reasonable. If it is too small in comparison to the size of the fee or the significance of the potential damages that could occur, a court may refuse to enforce it. Most important of all, the decision of this court demonstrates the value of seeking contract language where appropriate to limit the liability or the types of damages that can be recovered.

Chapter 22

Mold

22.1 Standards Needed for Mold Exposure, Testing and Remediation
22.2 Preventing Mold-Related Nondisclosure Claims
22.3 Court Rejects Employees' Claim against Employer for Fraudulent Concealment of Mold
22.4 Incident Reports are Held to be privileged
22.5 Homeowners Policy Unambiguously Excluded Coverage for Mold

22.1 Standards Needed for Mold Exposure, Testing and Remediation (Vol. 6, No. 3)

By: *Richard Zarandona & J. Kent Holland*

Currently, federal or state established standards of safety thresholds for mold exposure are are non-existent. Scientific agreement has not even been reached on whether mold (or various types of mold) are hazardous or injurious to health. Nor are there standards for mold testing and remediation.

The only authoritative standard we have found for examining, monitoring and remediating mold was written several thousand years ago. That's right. It's found in the Bible – Old Testament (or Pentateuch) at Leviticus 14:33-45.

With regard to mold or "spreading mildew," "The Lord said to Moses," that if an owner of a house sees mildew in his or her house this is what must be done:

"The owner of the house must go and tell the priest, 'I have seen something that looks like mildew in my house.' The priest is to order the house to be emptied before he goes in to examine the mildew, so that nothing in the house will be pronounced unclean. After this the priest is to go in and inspect the house. He is to examine the mildew on the walls, and if it has greenish or reddish depressions that appear to be deeper than the surface of the wall, the priest shall go out the doorway of the house and close it up for seven days. On the seventh day the priest shall return to inspect the house. If the mildew has spread on the walls, he is to order that the contaminated stones be torn out and thrown into an unclean place outside the town. He must have all the inside walls of the house scraped and the material that is scraped off dumped into an unclean place outside the town. Then they are to take other stones to replace these and take new clay and plaster the house. If the mildew reappears in the house after the stones have been torn out and the house scraped and plastered, the priest is to go and examine it and, if the mildew has spread in the house, it is a destructive mildew; the house is unclean. It must be torn down—its stones, timbers and all the plaster—and taken out of the town to an unclean place."

Today, in a world without standards, we are more likely to get less consistency and more unpleasant surprises, as described in the Texas newspaper report below. Texas has been one of the states with a skyrocketing number of reported cases of mold related claims by homeowners. According to an article in "The Dallas Morning News," (February 16, 2003 , Ed Timms reporting), "Across the state, Texans whose homes were gutted, or left unfinished, say they are casualties of botched mold remediation. Some fly-by-night companies, they say, have abandoned jobs after stripping the interior of houses to the studs.

Homeowners complain that mold remediators failed to properly contain mold in their homes, charged for services that were not provided, forged signatures, falsified claims and walked away with the jobs unfinished." The newspaper reports that one Dallas homeowner was billed three times the tax value of his home, and that numerous others were billed for cleanup costs that exceed their home's value. The article goes on with example after example of homeowners that have had terrible experiences with contractors they hired to remediate their homes of mold.

The Dallas Morning News reporter asks rhetorically, "What does it take to be a mold remediator?" and he answers, "While the state requires manicurists to complete 600 hours of instruction in an approved program, the training requirements for mold remediators are not quite so rigorous: *There are none.*" He then quotes Brenda Wells, a University of North Texas associate professor of insurance and director of the university's Financial Services Center as follows: "You've got to have a license in this state to kill bugs, but you don't have to have a license to tell an insurance company you need to spend $50,000 instead of $5 to clean up mold. I literally can take out a Yellow Page ad tomorrow that says I'm a mold remediator [and] I don't know the first thing about cleaning up mold."

The Mold Litigation Explosion: Mold is turning to gold for plaintiff's lawyers. An article in the American Bar Association (ABA) Journal (December 2001), interviews Alexander Robertson IV, an attorney in California that has earned "multimillions" for his firm handling toxic mold litigation. When asked how many cases he had, he answered, "Thousands—I don't want to count them." He explained, "The use of asbestos isn't occurring anymore, and most of the asbestos products were done away with. With mold, it's naturally occurring, and the supply is endless."

Another attorney quoted in the ABA Journal article, Guy Keith Vann, says, "I've learned these can be terrific cases from the plaintiff's perspective, in terms of the percentage of cases that turn into money vs. vases that don't." He obtained a $1million toxic-mold verdict that was sustained on appeal in the case of *New Haverford Partnership v. Elizabeth Stroot*, 772 A.2d 792 (Del.1999).

From reading through numerous articles in newspapers, magazines and journals, on the subject of the mold threat, it is apparent that there are a growing number of law suits and claims alleging that homes must be remediated or even destroyed because of a mold "problem." Yet, many of these same articles and papers quote extensively from medical specialists and other professionals who state as a general proposition that mold is not a problem for people other than those who have allergies, and that even then, it is not nearly the problem it is made out to be.

It has been estimated that over 10,000 cases related to mold have been filed in the United States . And the number seems to be growing exponentially. Construction contractors are brought into these cases on the theory that their defective means, methods and procedures of performing their work caused water leakage, condensation accumulation or other conditions that caused the growth of mold. Design professionals are being brought into these cases on various theories such as specifying improper materials for construction, failing to specify a design to effectively eliminate or minimize mold growth; and failure to adequately monitor or review construction during the construction phase of the project.

Suppliers of various materials such as wall and roofing materials and water pipes and couplings are being sued on theories such as their materials (a) failed to keep water from infiltrating a building, (b) failed to allow water to seep back out of a building, (c) were too easy a source for mold to grow; or (d) in the case of pipes and couplings, were too subject to leakage inside of wall cavities. As seen from the newspaper articles cited herein, the remediation firms that are subsequently hired to repair water damage and remediate mold are also subject to suit for faulty workmanship during remediation.

As reported in the *Corpus Christi Caller Times* (November 3, 2002 , Joy Victory reporting), "Mold claims filed with insurance companies have taken off in just the past few years. Farmers Insurance Group had 11 mold claims filed in 1999 and 10,813 in 2001. The average cost of cleaning up mold also grew eight times between 2000 and 2001, going from $17.09 per policyholder in the first part of 2000 to $147.68 in the second quarter of this year, according to the top three insurance companies in Texas ."

Connecting Mold to Adverse Health Effects: According to the *Corpus Christi Caller Times* article, the Texas Medical Association's Council of Scientific Affairs commissioned a literature review and found no reputable studies linking health problems to mold. They concluded that black mold only causes problems in people who are allergic to it. And for those who are allergic to the mold, the newspaper cited professional allergists for the conclusion that "The allergy symptoms probably will be no worse than a cat or dog dander allergy, causing symptoms such as congestion, sneezing and water eyes. In some people, it can cause asthma-like problems."

The supervisor of health hazard evaluation with the Wisconsin Department of Health and Family Service, William Otto, was quoted by the *Milwaukee Journal Sentinel* (December 13, 2002, Michele Derus reporting), as saying, "There's a lot of debate out there about what symptoms are from mold and what aren't. To us, the health effects are allergy-type symptoms. Maybe 20 percent of the population has some type of sensitivity. But at what level does someone get sick? That seems pretty much an individual thing." Otto was further quoted as saying, "That doesn't mean we can't address the problem. We emphasize getting at the moisture source, fixing it and getting rid of the contamination. More and more state health departments are following that philosophy."

But there are certainly numerous reported instances where individuals have asserted that their health was so significantly impacted by mold that they could no longer work or enjoy a good quality of life. In the *Detroit Free Press* (February 18, 2003, Patricia Anstett reporting), the story is told of a medical doctor who, after his building was renovated in 1999, started sneezing, suffering with itchy eyes, nosebleeds, shortness of breath, skin rashes and fatigue. Environmental health firms found "foul-smelling black mold substance blanketing his office—the result of several flooding incidents that were attributed to a construction flaw." Prior to the renovation, it is reported that he hadn't missed a day of work in 28 years, but due to his worsening in the months after the renovation, he hasn't worked since November 25, 2000 .

Courts in some jurisdictions have permitted testimony linking adverse health to mold exposure. See, for example, *Mondelli v. Kendel Homes Corp.*, 631 N.W.2d 846, 856 (Neb. 2001)(plaintiff alleged asthma –related symptoms from mold). Courts in other jurisdictions, however, have excluded testimony attempting to make the link between mold exposure and adverse health. See, *National Bank of Commerce v. Associated Milk Producers, Inc.*, 22 F.Supp.2d 942 (E.D. Ark. 1998)

Defective Construction is the Alleged Cause of Mold Growth: Some reasons offered for the recent increase in mold litigation include the following: (1) fast construction during the housing boom of the past several years has led to faulty workmanship; (2) also as a result of the fast pace of construction, lumber has been used before it sufficiently dried and materials have been permitted to lay uncovered and exposed in the weather before their use; (3) complex designs with multiple roof angles and gables, skylights, innovative angular and tiered wall systems, and the like have exacerbated the likelihood of leakage; (4) as buildings have become more air-tight, they may prevent materials that get wet either by condensation or leakage from drying out; and (5) building materials that are being used today are more susceptible to mold growth. The number of theories and potential defendants seems to grow almost as fast as mold itself.

As with the medical building renovation reported in the Detroit paper, it appears that many of the reported cases of mold problems began following renovation or remodeling of an existing house or office. In many other cases, the problem is alleged to be caused by faulty new construction that permitted leaks of water into the building from roofs or wall systems, or from interior water pipes. Other cases have alleged that condensation from HVAC systems has caused or contributed to the growth of mold.

In another article in *The Milwaukee Journal Sentinel* (December 29, 2002, Dan Benson reporting), the reporter describes the situation of a Michigan family that moved out of their house into a hotel for two months, and then into a rental home (all at insurance company expense) following a remodeling job on their home. They allege that faulty workmanship resulted in leaks and condensation around light fixtures, causing mold to grow on the roof sheathing, soffit area, and even in the basement. Now they are in a dispute with their homeowner's insurance company over repairs and damages.

A jury in one Texas case awarded judgment for a homeowner against her homeowners' insurance carrier in the amount of $32 million, on a claim property damage, bodily injury and mental anguish, all resulting from water damage and mold. In Florida, a jury awarded judgment over $11 million to Martin County Florida against a construction manager and its sureties for water damages and mold caused by construction defects. There was water infiltration through the exterior synthetic hard coat wall system and there were problems with the HVAC system. A significant part of the County's monetary damages arose from its decision to vacate the entire building while construction problems were corrected and remediation was carried out.

The Insurance Company's Perspective: Insurance premiums have historically been based upon actuarial information and underwriting guidelines that were in place before the massive mold litigation began. Consequently, the premiums collected have only covered risks and losses from the old bread-and-butter issues surrounding design and construction. Insurance companies that are defending against mold-related claims, and in some instances paying significant damages for mold-related claims, are being hurt financially because of insufficient collected premium to cover unforeseen claims on risks for which they charged no premium.

With new insurance policies for homeowners, and for design professionals and contractors that are involved in home design or construction, insurance companies are at significant risk unless they do something to limit their own exposure to loss resulting from claims alleging damages arising out of mold-related matters. For reasons such as those discussed in this newsletter, insurance companies currently underwriting policies such as these are finding it difficult, or even impossible, to evaluate the risk of potential mold-related damages and claims. Without actuarial data and scientific information to assist the underwriters, it is impossible to know how much premium is actually necessary to cover the potential risk. In the absence of standards for mold exposure and for mold testing and remediation, it is hard to see how this situation will change anytime in the near future.

In the meantime, some insurance companies are devising ways to provide limited mold coverage to some of their insureds, while at the same time protecting the insurance company against the financial disaster that could ensue from an avalanche of mold claims. For example, it may be possible to endorse policies to provide a sub-limit of coverage for mold. A design professional liability policy or contractor's general liability policy, for example might include an endorsement stating that damages related to mold shall be limited to a dollar amount less than the full policy limit. Another approach is to limit coverage to property damage only, and to specifically exclude bodily injury claims. It is likely that for the foreseeable future, decisions concerning the terms, conditions and amounts (if any) for granting mold-related coverage in new policies will be made by insurance carriers on a case-by-case basis.

Need for Standards: It is clear that there is a need for standards related to mold. Until standards are established, it will be hard for insurers to cover design professionals and contractors for this exposure. Many of these firms may have to operate without coverage or with limited coverage that may expose them to serious financial risk. Standards are needed now, and they need to address, at a minimum, licensing, and certification of mold evaluators and mitigators. Standard testing methods are needed to enable accurate and consistent analysis of the potential problems. Establishing criteria and methodology for clean-up must be accomplished in order to determine a reasonable standard of care for designing and carrying out mold mitigation efforts.

Finally, more research needs to be done by the medical community to develop quantitative exposure guidelines and causal links to health effects. A cooperative effort between local, state and federal agencies is needed—and needed soon—so that we can move ahead from the biblical standards for dealing with mold in earthen houses to the standards required by a modern society.

Acknowledgment: This article was originally written for publication in Environmental Risk Briefings, a newsletter published by Arch Insurance Company. The Briefings are available at the Arch Insurance website at http://archinsurance.com/aigu_mkt_environ_index.htm.

About the Authors: Richard Zarandona is a Senior Vice President for the Arch Insurance Group where he directs and manages the Environmental and Design Professional operations. Mr. Zarandona holds and MBA in Finance from the Silberman School of Business. He also has a Masters Degree in engineering and is a licensed professional engineer in numerous states. Mr. Zarandona has spent more than 25 years working in the environmental and insurance fields. He has provided expert testimony for environmental projects and has managed Phase I and Phase II site assessments, site remediations and numerous Federal and State clean up actions and studies.

Kent Holland is Risk Management Services Consultant for the Environmental and Design Professional Unit of Arch Insurance Group, Inc., and is an attorney with the law firm of Wickwire Gavin, P.C, as well as publisher of *ConstructionRisk.com Report*.

22.2 Preventing Mold-Related Nondisclosure Claims
(Vol. 5, No. 8)

By: *Gordon Rees, L.L.P.*

Property owners and managers are increasingly faced with claims due to actual and potential indoor mold problems, raising the dilemma of what needs to be disclosed when leasing and selling properties. Although California's Toxic Mold Protection Act, signed into law on October 1, 2001, includes mold disclosure requirements for building owners, the requirements do not go into effect unless and until a California Department of Health Services taskforce first determines objective "permissible exposure levels," which has not yet been done.

The Department of Health Services cites a lack of funding to explain its inability to convene the taskforce and address the threshold question of whether permissible exposure levels for indoor molds are feasible. Despite the lack of regulations mandating mold-related disclosures, the California Association of Realtors recently added a specific question to its standard CAR disclosure form to inquire about the presence of mold. Moreover, given the media attention and substantial jury awards, the number of real estate non-disclosure actions involving mold has increased substantially.

Under California law, buyers and tenants of commercial property are traditionally deemed to be sophisticated and able to freely negotiate contracts with owners. As long as there are no material or fraudulent misrepresentations, commercial owners generally only need to be concerned with disclosing any "known material facts" that affect the value or desirability of the property, and there is no obligation to repair defective conditions.

Residential sellers and landlords generally need to be more wary of rendering complete written disclosures because even if they have no actual knowledge, residential owners are also required to disclose any condition that they "should have known." [*See, California Civil Code §§ 1102*, et seq. for more information.] Therefore, when in doubt, disclose, disclose, disclose!

For example, Owner X wants to sell his single family residence and is not aware of having any mold problems during the ten years he occupied the home. However, five years earlier, a pipe had burst; the kitchen had flooded; and water had leaked into the crawl space and other parts of the home. Professional contractors repaired the damage, and Owner X had no further problems. Now that Owner X is selling his home, should he disclose the prior flood? <u>Yes</u>. Should he disclose the repairs and subsequent inspections that were done? <u>Yes</u>. Should he disclose any reports he obtained? <u>Yes</u>. Should he even disclose the minor flooring defect he discovered when repairing the flood damage? <u>Yes</u>. And in the mold context, these same answers would also generally apply to a commercial property transaction.

The tricky question is whether to disclose the "potential" for mold due to the extensive water intrusion into the crawl space, wall cavities, etc. Although the parties in a *commercial* setting typically would be deemed to be equally aware that water intrusion can lead to mold growth, buyers and tenants may nonetheless claim ignorance and pursue a claim for non-disclosure. In the *residential* context, plaintiffs often claim ignorance as to the connection between water intrusion and indoor mold growth. Such plaintiffs may sue based on the lack of disclosure of water intrusion and/or mold and seek damages such as remediation and repair of personal and real property, bodily injuries, fraudulent concealment and misrepresentation claims, "stigma" and resultant "diminution of the home's value and/or breach of contract. These cases can be quite expensive to defend, and owners and/or realtors may not have applicable or adequate insurance to cover the costs of defense or judgments. Therefore, it is typically advisable to err on the side of disclosing all past water intrusion and mold-related events as well as the investigation and repairs made, if any, when selling either commercial or residential property.

22.3 Court Rejects Employees' Claim against Employer for Fraudulent Concealment of Mold
(Vol. 5, No. 8)

By: *Gordon Rees, L.L.P.*

The California Appellate Court (2nd Dist.) recently affirmed the trial court's summary adjudication of an employee's tort action brought against her employer. *Jensen v. Amgen, Inc.* (2003) 105 Cal. App. 4th 1322. An employee injured during the course of employment is generally limited to remedies available under the Worker's Compensation Act. However, there is a narrow exception to this exclusivity rule where the employee's injury is aggravated by the employer's fraudulent concealment of the existence of the injury and its connection with the employment.

The narrow exception was first articulated in the asbestos context. The employee alleged that his employer learned that he had an asbestos-related disease through routine screening and fraudulently concealed this condition from him, thereby preventing him from receiving treatment for the disease and inducing him to continue working under hazardous conditions. *John Mansville Products Corp v. Sup. Ct.* (1980) 27 Cal.3d 465.

Here, plaintiff complained to a company nurse of having sinus headaches, skin rashes and fatigue. At the time, plaintiff attributed her symptoms to allergies to the laboratory animals present and was transferred. A few months later, a mushroom was discovered in the building. Subsequent air testing revealed the presence of "toxic mold" but in concentrations that were lower indoors than outdoors. The employer informed building occupants of the mold and removed and repaired the water intrusion and mold. Plaintiff subsequently learned that mold had also been discovered and cleaned five years prior in the air delivery system of the building.

Plaintiff took a medical leave of absence and filed a lawsuit for fraudulent concealment against her employer alleging that her injuries were mold- related and her employer knew of the presence of the "toxic mold" in the building, knew that plaintiff's symptoms were related to the "toxic mold" and concealed this information from her.

The court rejected plaintiff's arguments and confirmed the extremely narrow exception to the worker's compensation exclusivity rule. The court held that the threshold issue was whether the employer knew of plaintiff's symptoms before she knew of her own symptoms. The court noted that the exceptions to the worker's compensation exclusion are intended to be extremely limited. The court also noted that plaintiff was the first person to associate her symptoms with mold in the building. There was no evidence that the employer was aware of any such connection. Rather, the employee had told the employer that her symptoms were caused by animal allergies. Moreover, the employer relocated plaintiff after learning of her allegedly building-related symptoms. Therefore, plaintiff failed to establish the elements of fraudulent concealment and the court affirmed the summary judgment in favor of the employer.

22.4 Incident Reports are Held to be privileged
(Vol. 5, No. 8)

By: *Gordon Rees, L.L.P.*

A California Court of Appeals held that incident reports are protected by the attorney-client privilege when certain criteria are met. *Scripps Health v. Superior Court (Reynolds)*, 2003 D.A.R. 6059 (filed June 6, 2003) This case is particularly instructive to companies, such as property owners and managers, which regularly confront mold claims and have established protocols requiring the completion of report forms when water intrusion complaints are received.

This case sets forth criteria for maintaining the privileged nature of this material.

In this wrongful death case against a hospital, plaintiffs filed a motion compelling production of internal incident reports. In opposition to the motion to compel, the hospital submitted evidence that (1) the reports were marked confidential, (2) the reports were used by the hospital's attorneys to assess internal risks and create a claims profile and (3) access to the reports was limited to risk managers, in-house or outside counsel and third party claims administrators. To the extent that information from the reports was needed by other departments at the hospital, it was extracted from the documents and then circulated in another format.

The trial court granted plaintiffs' motion to compel, but the Court of Appeals reversed, holding that the reports were protected by the attorney-client privilege. The court reasoned that the existence of the privilege depends more upon the intended and actual use of the document than its contents. The court emphasized the significance of forwarding the reports to the Legal Department, risk managers or outside counsel. If copies are kept only by the initiating department, there is a stronger argument that the primary purpose of the reports is not communication with attorneys, but customer service or other administrative purposes.

Significantly, the court did not limit its holding to hospital settings, but rather on application of the attorney-client privilege to the corporate setting in general. Accordingly, any company that creates confidential records involving incidents which may result in litigation may be able to classify the documents under attorney-client privilege if the above-listed criteria is followed.

About the Author: Case Notes 22.2, 22.3, and 22.4 were originally published by the law firm of Gordon & Rees L.L.P.

in the July/August 2003 issue (Vol.3, No. 3) of their newsletter entitled, *"Mold... Matters!"* The attorneys contributing to the articles include Mike Pietrykowski, Linda Moin, Traci Lagasse and Molly McKay. The San Francisco office address is Gordon & Rees, 275 Battery St., Suite 2000, San Francisco, CA 94111; (415) 986-5900. The firm also has offices in San Diego, Los Angeles, Sacramento, Orange County, Portland, OR, and Las Vegas, NV. For more information on the article you may also contact the authors by e-mail at: mpietrykowski@gordonrees.com. The firms website has additional information and articles. See http://www.gordonrees.com.

22.5 Homeowners Policy Unambiguously Excluded Coverage for Mold

(IRMI Expert Commentary, K. Holland - October 2004

Where mold damage allegedly arose out of shoddy roofing work, a homeowner attempted to recover under its homeowners policy for the mold as an "ensuing loss" despite a provision of the policy specifically excluding coverage for mold contamination. In reviewing the policy language the court concluded that the mold exclusion was clear and unambiguous, and therefore, losses caused by mold were not covered.

In *Brick v Lexington Insurance Company* (No. ATL–L–1285–03 (April 2, 2004—Superior Ct of NJ)), a homeowner sought to recover property damage under its homeowners policy allegedly caused by poor work performed by a roofing contractor. There was a dispute as to whether the damage was due to vandalism or poor workmanship. The policy had an exclusion for damage or losses resulting from poor workmanship. The plaintiff argued that the damage was caused by vandalism by the contractor when it walked off the job. In addition, the plaintiff argued that the policy did not exclude coverage for mold contamination which was an "ensuing loss" to loss caused by vandalism.

The Lexington policy under Coverage A (Dwelling) and B (Other Structures) provided in pertinent part as follows:

> We insure against risk of direct loss to property described in Coverages A and B only if that loss is a physical loss to property. We do not insure, however, for loss: 2. caused by: e.(3) smog, rust or other corrosion, fungus, mold, wet or dry rot; [or] 3. Excluded under Section 1—Exclusions.

The policy also stated: "Under items 1 and 2, any ensuing loss to property described in Coverages A and B not excluded for excepted in this policy is covered."

In analyzing whether Lexington should have been granted a motion for partial summary judgment, the court reviewed the language of the policy and stated that there is no ambiguity about exclusion of mold so that "losses caused by mold are not covered to the structure." The court went on to explain, however, that the dispute focuses on the "ensuing loss" coverage of the policy. The plaintiff argued that while losses directly caused by mold are not covered, that where the loss is caused by vandalism which would be a covered loss and where mold ensues from vandalism, then mold damages would be covered. The court concluded that this position was not supported by case law or any reasonable interpretation of the policy. Decisions by courts in several other states were reviewed and quoted by this court in explaining its decision.

One decision reviewed by the court was from the case of Fiess v State Farm (SD Tex), in which the Texas court held that "for coverage to be restored via the ensuing loss clause an otherwise covered loss must result or ensue from the excluded loss." The Texas court held that the "ensuing loss" language in the Fiess policy:

> means mold itself because it is specifically excluded is never covered but if mold caused a "covered loss," then that "covered loss" would be covered under the "ensuing loss" language.

The New Jersey case in Lexington referenced four other decisions from various state courts and concluded that courts hold that:

> the ensuing loss provision does not reinsert coverage for excluded losses, but affirms coverage for secondary losses ultimately caused by excluded perils.

The ensuring loss provision means that if one of the specified uncovered events takes place, an ensuing loss that is otherwise covered remains covered but the uncovered event itself is never covered. Applying that reasoning in this case, the New Jersey court concluded that mold damage to the residence would not be covered under the terms of the policy or under the ensuing loss exception to the mold exclusion. To hold otherwise, said the court, "would render mold exclusions basically meaningless since mold always is caused by other events and therefore is always an 'ensuing loss.'"

Comment

Although this particular decision dealt with a homeowners policy, the principles of policy interpretation explained by the court apply equally in the context of contractors' general liability policies and design professional errors and omissions policies. Where there are clearly stated exclusions for damages arising out of mold growth or contamination, it should be anticipated that exceptions to the exclusions will be narrowly interpreted so as not to make meaningless the plain intent of "mold exclusions." Contractors and design professionals that are concerned about coverage for mold may pursue various options for obtaining coverage by adding endorsements to their policies to specifically cover damages arising out of mold.

There are a number of ways that insurance carriers may be willing to provide mold coverage, including for example, a separate lower sublimit for mold or a higher deductible for mold. Other endorsements may provide mold coverage provided it does not result from poor workmanship or poor maintenance of a building. Whether mold coverage will be made available to a specific account will also depend on the nature of the construction (residential or commercial) and the location of the construction, and a number of other factors that insurer include in their underwriting guidelines.

Chapter 23

Site Safety

23.1 Store Owner not Liable for Injuries Sustained by HVAC Contractor's Employee
23.2 Engineer Had No Duty to Warn General Contractor's Employee of Danger
23.3 No Liability Under New York Labor Law for Project Owner and Lender where Worker's Injuries Attributed Solely to his Own Fault

23.1 Store Owner not Liable for Injuries Sustained by HVAC Contractor's Employee (Vol. 7, No. 5)

Where a contractor's employee was injured by falling from a ladder it borrowed from the owner of the location where he was repairing an HVAC system, the court held the owner was not legally responsible for the individual's injuries because the availability of the ladder was at most a favor to the person doing the work. This was nothing more than a mere gratuity, for which the owner owed no duty to the individual.

In the case of *Semler v. Sears, Roebuck and Company*, 268 Neb. 857, 689 N.W. 2d, 2004, Lawrence Semler, an employee of The Waldinger Corporation was dispatched by his employer to the Sears store to repair the heating unit. Upon his arrival at the store, Semler noticed a ladder leaning against the heating unit, and he climbed it in order to take a look at the heating problem. He came back down, went to his truck to get an electric meter, then returned to the heating unit and climbed the ladder a second time. While he was on the ladder, the bottom slipped out across the floor, causing him to fall to the ground.

Semler testified at trial that he believed the ladder's lack of rubber shoes caused the ladder to slip out on the concrete floor. He also testified that he "most likely" adjusted the ladder before climbing it, but that he did not notice until after the accident whether the "shoes" on the ladder had rubber on them. He further testified that his employer, Waldinger, provided all the tools needed for the job, including an extension ladder which was on the roof of the truck he drove to the job. He said he chose not to use that ladder, however, because there was already a ladder on the premises. The trial court weighed conflicting testimony and concluded that Sears employees did not retrieve or move the ladder for Semler to use, but that Semler, himself, made an independent decision to use the ladder without any involvement by Sears.

The appellate court stated that the issue for consideration was not whether Sears had retained control over Semler's work. Nor did the court believe there was a legal issue of whether or not Sears had a non-delegable duty to provide a safe workplace for Semler. In fact, Semler did not argue that Sears had retained control over an independent contractor who had caused him harm. And he did not argue that Sears had vicarious liability for actions of an independent contractor. Nor did he argue that his injuries were due to Sears' failure to protect him form a condition or activity existing upon its land. Instead, Semler sued Sears for direct negligence in supplying a defective ladder for his use on its premises.

The trial court found no legal basis for the claim and granted a summary judgment in favor of Sears against Semler. On appeal, the appellate court affirmed the summary judgment, finding that, "At most, the presence of the ladder leaning against the unit could be viewed as a 'favor to the person [Semler] doing the work." As such, "its availability would be nothing more than a mere 'gratuity." Consequently, under the Restatement of Torts, Sec. 392, as cited by the court, Sears owed no duty to Semler.

23.2 Engineer Had No Duty to Warn General Contractor's Employee of Danger (Vol. 6, No. 2)

An engineer was not liable for injuries sustained by an employee of a construction contractor where the engineer's contract did not make the engineer responsible for site safety and where the engineer did not take any action in the field to supervise or control the contractor's work or otherwise involve itself in the contractor's safety practices.

In the case of *Hobson v. Waggoner Engineering, Inc.,* 2003 WL 21789396 (Miss. App. 2003), suit was brought against an engineer by the estate of a deceased worker employed by a subcontractor to the general construction contractor. The worker died by drowning in a lagoon that was being built for a wastewater treatment plant, designed by the engineer. The engineer's role during construction was limited to monitoring the progress of the construction work for general compliance with the plans and specifications. Its contract did not give the engineer authority to stop work or issue change orders but required it to make recommendations for approval by the project owner. In contrast, the construction contract gave the contractor full responsibility for site safety.

Allegations by the worker's estate included that the engineer defectively designed the lagoon by making it's liner sides too slick and its incline too steep to be climbed out of by a worker. The estate also argued that the engineer violated a duty to warn the worker that the steep slides and slick surface posed a danger.

The court held the engineer had no duty to the construction worker because the contract gave the engineer no site safety responsibility and the engineer did not act outside its contractual authority to take on any site safety responsibility. On the issue of the alleged defective design, the court held the engineer could not be found liable in the absence of evidence via expert testimony (which the plaintiff failed to present) that the slope and slickness of the liner was contrary to the standard of care. Although the plaintiff's expert testified that there was insufficient traction for the worker to climb out of the lagoon, he did not testify that any alternative design would have been available or that the engineer did not comply with the standard of care. Without such expert testimony, the issue could not even go to a jury for consideration and a summary judgment was appropriately granted in favor of the engineer by the trial court.

Comment: This decision reaches a reasonable conclusion concerning site safety responsibility by honoring the intent of the contracting parties and recognizing the appropriately restricted role of design professionals during construction. Another case that we reported on last year was the case of *Herczeg v. Hampton Township Municipal Authority and Bankson Engineers*, 766 A.2d 866 (2001). In that case the Pennsylvania Supreme Court held: "We are not persuaded that the rationales expressed in these cases [such as *Carvalho*] warrants the establishment of a new rule of law fastening liability based strictly upon an assertion of actual knowledge of unsafe work-site conditions." "We reject any notion that a duty arises based solely upon an engineer's actual knowledge of dangerous conditions.... If someone is under no legal duty to act, it matters not whether that person is actually aware of a dangerous condition.... Conversely, if someone by contract or course of conduct has undertaken the responsibility for worker safety that person may still be liable even in the absence of actual knowledge of the dangerous condition if they should have known of the condition."

The factors which would appear to be relevant in any case where an attempt is made to expand a design professional's liability beyond the specific provisions of its contract with the owner include the following:

(1) actual supervision and control of the work;
(2) retention of the right to supervise and control;
(3) constant participation in ongoing activities at the construction site;
(4) supervision and coordination of subcontractors;
(5) assumption of responsibility for safety practices;
(6) authority to issue change orders; and
(7) the right to stop the work.

These same factors may be used by courts in evaluating whether the project owner retained sufficient control over the site to be held responsible for site safety. Where an owner does not retain any control over the manner in which the work was performed, and it has no actual knowledge of the danger or condition that resulted in a laborer being injured, it generally has no liability according to the cases reported in this course.

General contractors should also be aware of these factors that are used by the courts, because liability of a general contractor for injuries to employees of subcontractors may be limited to those situations where the general contractor has retained control over the operative details of the subcontractor's work.

23.3 No Liability Under New York Labor Law for Project Owner and Lender where Worker's Injuries Attributed Solely to his Own Fault
(Vol. 6, No. 1)

The strict liability imposed by Section 240 of the New Labor Law appears to have been somewhat clarified in the decision of *Rupert Blake v. Neighborhood Housing Services of New York City, Inc.,* 2003 NY Slip Opinion 19690, Court of Appeals (Dec. 2003). In this case, the plaintiff was an individual that operated his own contracting company. He was working alone on a renovation job at a two-family house when the ladder on which he was standing collapsed, resulting in the plaintiff breaking his ankle.

The plaintiff filed suit against Neighborhood Housing Services of New York City (NHS), a not-for-profit lender, which had provided low-interest financing to facilitate the project. NHS had nothing to do with construction decisions or supervision. Its role was limited to dispatching a rehabilitation specialist to the premises to assess the scope of the work and the amount of the loan. NHS prepared a work estimate and gave the homeowner a list of contractors, from which she chose plaintiff.

Plaintiff contended that NHS was strictly liable under New York Labor Law, section 240(1) as a statutory agent under the section for having failed to provide a proper workplace and mandated safety equipment. At trail, a jury found against the plaintiff on the facts presented. The trial court denied plaintiff's motion to vacate the jury's verdict and direct one in his favor. The Appellate Division affirmed, stating that "a factual issue was posed as to whether plaintiff's injury was caused by some inadequacy of the ladder or was solely attributable to the manner in which plaintiff used the ladder" and that there were no grounds to disturb the jury's factual determinations. On appeal, the plaintiff argued Labor

Law § 240 (1) is a strict (or absolute) liability statute and that the court should have set aside the jury's verdict.

The question for the court was whether a plaintiff who was injured while using a ladder may prevail in a Labor Law § 240 (1) action even when a jury finds that the ladder was so constructed and operated as to give him proper protection and he was the sole cause of his injury.

The court stated that in deciding the appeal, it is necessary to address the concept of strict (or absolute) liability and the predicates for its application under Labor Law § 240 (1). The court explained that it has in numerous previous decisions "steadfastly held that contributory negligence will not exonerate a defendant who has violated the statute and proximately caused a plaintiff's injury." But the Court states that: "At no time, however, did the Court or the Legislature ever suggest that a defendant should be treated as an insurer after having furnished a safe workplace. The point of Labor Law § 240 (1) is to compel contractors and owners to comply with the law, not to penalize them when they have done so."

"Plaintiff asserts, in essence, that despite the jury's findings he is entitled to recover because Labor Law § 240 (1) provides for strict (or absolute) liability. In addressing this contention, we note that the words strict or absolute liability do not appear in Labor Law § 240 (1) or any of its predecessors. Indeed, it was the Court — and not the Legislature — that began to use this terminology in 1923 (under an earlier version of the statute [*see* L 1921, ch 50]), holding that employers had an "absolute duty" to furnish safe scaffolding and would be liable when they failed to do so and injury resulted."

The Court states that it has always stressed two points in applying the doctrine of strict (or absolute) liability under section 240 (1) of the Labor Law. "First, that liability is contingent on a statutory violation and proximate cause.... [and] second, that when those elements are established, contributory negligence cannot defeat the plaintiff's claim." It is imperative, therefore, to recognize that the phrase "strict (or absolute) liability" in the Labor Law § 240 (1) context is different from the use of the term elsewhere, says the Court. Thus, says the Court, "an accident alone does not establish a Labor Law § 240 (1) violation or causation."

The court rejected the plaintiff's argument that "he is entitled to recover in the face of a record that shows no violation and that reveals that he was entirely responsible for his own injuries." To impose liability for a ladder injury even though all the proper safety precautions were met would not further the Legislature's purpose, concluded the court. The court stated: "If liability were to attach even though the proper safety devices were entirely sound and in place, the Legislature would have simply said so, or made owners and contractors into insurers. Instead, the Legislature has enacted no-fault workers' compensation to address workplace injuries where, as here, the worker is entirely at fault and there has been no Labor Law violation shown."

An additional reason the Court gave for rejecting the Plaintiff's suit and holding NHS could not be liable under the New York Labor Law was that NHS was not itself a contractor or project owner, or an agent of a project owner. Authority would have had to be given to a NHS by the owner to supervise or control the work in order for NHS to be deemed an agent for purposes of Labor Law section 240 (1). The court states: "Although defendant [NHS] here coordinated home repair work, it did not involve itself with the details of how individual contractors would perform their jobs. Instead, NHS acted as a lender: it is a non-profit organization that provides low-interest loans. The homeowner retained primary control over decisions on how the renovation project would proceed. NHS did not supervise the contractor; it never instructed workers on how to undertake repairs, and it took only a de minimis role in ensuring that the contractor would complete the financed repairs.

Chapter 24

Standard of Care

24.1 Architect Required to Review Adequacy of Engineer's Structural Report Before Proceeding with its Design Services
24.2 Summary Judgment against Plaintiff who Failed to Provide Factual Support that she could Meet Burden of Proof of Negligent Design
24.3 Design-Build Engineer Held Liable for Negligence

24.1 Architect Required to Review Adequacy of Engineer's Structural Report Before Proceeding with its Design Services (Vol. 7, No. 4)

A material factual dispute existed regarding whether an architect met the requisite standard of care in reviewing an engineering report prior to its client's purchase of property and prior to the architect proceeding with plans for renovation of the building. Contractual language stating that the architect could rely on independent reports prepared for the owner by other firms did not relieve the architect of its responsibility to exercise reasonable care to review the engineering report and request additional inspection if it deemed it necessary for conducting its services.

A purchaser, Kerry, retained an architect, Angus-Young, to study potential sites for a business facility. The architect noted that the "structural stability of the 'over the river' building was unknown at this time," and consequently recommended that Kerry obtain further inspection by a structural engineer. Kerry obtained such an inspection by Rust Environment & Infrastructure, which later became Earth Tech Environment & Infrastructure.

Based upon visual inspection, Rust found the building to be in good structural condition. Its report concluded that "at this time, the piers and caissons are not considered a significant safety concern. Kerry went forward with the purchase of the building and a retained Angus-Young to do the architectural services for renovating the building. As work began, the flooring was removed and it was discovered that one corner of the building was three and three-quarters inches lower than the rest of the floor.

Further engineering by another firm determined that the reason for the problem was that the building rested on timber piles below the water line instead of on concrete piles and caisson as had been thought. Much time and expense were involved in repairing the foundation before renovation could begin, and Kerry sought to recover that cost from Angus-Young and from Rust Environment because it asserts it wouldn't have purchased the property and incurred the additional cost if the true condition had been reported beforehand.

Kerry's theory against the Architect was that it breached its contract, committed negligence and made negligent misrepresentations by failing to properly review the Rust report and failing to properly determine the renovation project requirements and costs. Rust also filed a cross-claim against Angus-Young, for contribution and indemnification in the event it lost to Kerry. The trial judge ruled that Angus-Young was contractually entitled to rely on the Rust report which had been independently obtained by Kerry and furnished to Angus-Young, and it therefore granted summary judgment to the architect.

This was reversed on appeal, with the appellate court finding that the contract language did not absolve Angus-Young "from its duty to exercise 'due architectural care' in performing services for Kerry—that the duty pre-existed the parties' contract and attached itself to all of the activities Angus-Young performed in fulfilling its contract with Kerry relating to the renovation project."

The architect contended the scope of its services was limited and that evaluation of the adequacy of the Rust report was not within its scope of services. It also argued that the list of "optional additional services" that Kerry did not opt for must be considered in limiting the scope. Included in these optional services that Angus-Young was not required to perform were "providing planning surveys, site evaluations or comparative studies of prospective sites," "providing services to investigate existing conditions or facilities," "providing services to verify the accuracy of drawings or other information furnished by the Owner," and "providing services in connection with the work of ... separate consultants retained by the Owner."

The contract also specifically required Kerry, and not Angus-Young, to furnish "surveys describing physical characteristics..." And, finally, the contract provided that as to "services, information, surveys and reports" furnished by Kerry, Angus-Young was "entitled to rely upon the accuracy and completeness thereof."

Based on the above-quoted provisions of the contract, Angus-Young argued it had no responsibility to perform its own investigation and that it accepted the Rust report as furnished to it by Kerry without further inquiry or investigation. The court rejected this argument. It found that

Kerry wasn't arguing that Angus-Young was required to do its own structural inspection but rather that Angus-Young was required to exercise due care so as to recognize the inadequacy of the Rust report to serve as a basis for its preparation of architectural plans for the renovation of the building. In particular, Kerry argued that because the Rust report didn't say anything about the foundation below the water line and was also silent as to whether underwater features had been inspected, Angus-Young was required by the standard of care to call for a more detailed inspection before beginning to perform its own services.

The court particularly keyed in on language in the contract that permitted Angus-Young to retain consultants to assist it in performing its duties. It further focused on language of the contract that obligated Kerry to "furnish the services of other consultants when such services are reasonably required by the scope of the Project and are requested by the Architect." The fact that Angus-Young did not request additional services here was deemed important by the court.

For these reasons, the court reversed the summary judgment and remanded the case to the trial court for trial on the merits of the complaint. *Kerry Inc. v. Angus-Young Associates, Inc.*, 694 N.W.2d. 407 (Wis. App., 2005).

24.2 Summary Judgment against Plaintiff who Failed to Provide Factual Support that she could Meet Burden of Proof of Negligent Design (Vol. 7, No. 1)

When a bicyclist crossing a small bridge, slid on loose gravel, hit and flipped over a guard rail, and suffered injuries when she fell about twenty feet into a dry creek bed, she sued the local county as well as all engineers who had anything to do with the design and construction of the bridge. The trial court granted, and the appellate court affirmed, motions for summary judgment against the plaintiff because she failed to provide factual support sufficient to establish that she would be able to satisfy her evidentiary burden of proof at a trial.

In the case of *Boland v. West Feliciana Parish*, (2003 CA 1297, Docket No. 17,142, June 25, 2004), the court stated that Louisiana courts have adopted a duty-risk analysis in determining whether to impose liability for negligence. A plaintiff must prove the following five elements: (1) the defendant had a duty to meet a specific standard of care; (2) the defendant failed to meet the standard of care; (3) the defendant's failure to meet the standard was the cause-in-fact of the plaintiff's injuries; (4) the defendant's failure to meet the standard of care was a legal cause of the injuries; and (5) the extent of actual damages.

Duty, says the court, is a question of law, and the legal question of the existence of a duty is appropriately decided by a court through summary judgment rather than by a jury at trial where there is no factual dispute and no credibility determinations required to be made. Plaintiff's claims against the engineer (Decoteau) were based on his position as a consulting engineer before and during construction of the bridge. She asserted that the West Feliciana Parish (hereinafter referred to as "county") used his plans for the bridge and sought his advice during construction. Decoteau, however, claimed he had a very limited role, advising the county only on specific aspects of the bridge project, including reviewing bridge plans prepared by another engineer, helping the county obtain a third construction bid, and providing load limit calculations on the bridge after it was built. He supported his position with affidavits and deposition testimony of himself and others. He said he did no measurements of guard rail heights or other aspects of the bridge not affecting load limits. Further, he stated the county didn't ask him to do anything concerning construction other than a structural examination to determine load limits after construction was completed.

In view of the pleadings, affidavits and depositions, the appellate court agreed with the trial court that the plaintiff did not establish that any of the engineer's duties affected the bridge conditions that allegedly caused her accident. For this reason, summary judgment was appropriately granted, said the court.

The court next reviewed whether summary judgment was appropriately granted to the county. The county argued that the plaintiff had failed to present sufficient factual support to establish the bridge was defective due to a condition that created an unreasonable risk of harm and that the parish knew of the risk. Her primary contentions concerning hazardous conditions involved the surface material and the height of the guard rails. She claimed that the crushed limestone aggregate presented an uneven and unstable surface that caused her bike to skid, and the county should have overlaid the material with a hard surface. She also claimed the railing was too low, in part because of design error and in part because of a build-up of gravel next to it.

As support for the county's claim that the bridge surface and railings were not defective and that the county had no knowledge of risk, the county submitted excerpts of several depositions, meeting minutes, AASHTO standards, and two affidavits by a professional engineer. Evidence presented showed that the constructed railing height satisfied the AASHTO standards, even if the county is correct that the standards don't apply to this small road and bridge. Affidavits and deposition testimony of a professional engineer stated that the crushed limestone aggregate surface was not unstable and that it was suitable for this type of bridge in this type of service.

Based on the evidence, the court concluded that the plaintiff failed to present factual support that the bridge was defective due to a condition that crated an unreasonable risk of harm and that the county had actual or constructive knowledge of any risk associated with it. Moreover, found the court, an affidavit submitted by a professional engineer on behalf of the plaintiff stated that it was the plaintiff's attempt to make a "sharp corrective turn" to stop her skid that "flipped the bicycle over toward the railing and created the centrifugal force sufficient to carry her over it, not the skid or some fulcrum effect created by the height of the railing itself and contact with it." For these reasons, the court affirmed summary judgment in favor of both the engineer and the county.

Comment: This case demonstrates the importance of aggressively pursuing motions for summary judgment based upon facts that can be discovered prior to trail. In this case, the court determined that since the plaintiff had failed to present a sufficient factual and legal case for a jury to find the defendants liable, it would be inappropriate for the court to do anything but grant summary judgment for the defendants.

Without regard or reference to the facts of this particular case (of which I have no personal knowledge), it's encouraging to see decisions where judges exercise their judicial discernment to terminate litigation against parties where the pre-trial pleadings and evidence do not substantiate including the parties in the lawsuits. Not every accident is the result of someone else's negligence. Some accidents just happen – without fault. That's unfortunate – but it's a fact of life. Tort reform is badly needed on the federal and state level to correct our legal system that seems to encourage plaintiff attorneys and their clients to turn every accident in to a tort (negligence) claim against others (preferably others who have large insurance policies).

24.3 Design-Build Engineer Held Liable for Negligence

(IRMI Expert Commentary, K. Holland - July 2001)

A complex court decision holds that a developer is entitled to recover its losses from the design-builder who employed the engineer and constructed the failed projects. Where a developer purchased land and constructed hydroelectric facilities in reliance on an engineer's analysis of the financial viability of proposed hydroelectric projects (including an assessment of the energy to be generated by each project as well as the costs of construction), it was entitled to recover its losses for the financially disastrous projects from the design-builder that employed the engineer and constructed the projects.

One Man—Multiple Hats

In *Hydro Investors, Inc. v Trafalgar Power, Inc.*, 227 F3d 8 (2nd Cir 2000), an engineer by the name of Neal Dunlevy wore multiple hats in his dealings with Trafalgar Power, Inc. (TPI), the entity that developed the plants. Mr. Dunlevy was an engineer with Stetson-Harza and also the principal and sole shareholder of Hydro-Investors, Inc. (HII). He approached TPI with the suggestion that hydroelectric plants could be profitably developed in upstate New York. He also suggested that TPI provide the capital to develop the plants and that he, Dunlevy, would contribute the hydroelectric expertise either through Stetson-Harza or HII. As a result of these discussions TPI retained Stetson-Harza and HII to perform work associated with TPI's licensing and possible development of six plants.

TPI entered into a contract with Dunlevy, through HII, calling for TPI to provide the capital while HII would identify sites to be developed. Dunlevy's activities for TPI were not limited to his work thought HII, but included a role as chief engineer of the Water Resources Department at Stetson-Harza.

As part of its effort to obtain project financing, TPI asked Stetson-Harza to prepare an analysis that could be used by potential lenders to evaluate whether to invest in the projects. Included in this analysis was an assessment of projected energy output.

The plants were constructed and, as stated by the court, "have proven to be financial disasters," the primary problems being costs overruns and inadequate energy production. At trial, a jury found Stetson-Harza and Dunlevy jointly and severally liable in the amount of $7.6 million for damages arising out of engineering malpractice.

Negligence—Proximate Cause of Damages

On appeal, Stetson-Harza and Dunlevy argued the jury's findings of engineering malpractice were unsustainable. They asserted that TPI failed to prove that the alleged acts of malpractice were the proximate cause of its injury, since "negligent measurements of the flow and head at Forestport and Ogdensburg did not cause those plants to generate less energy." Instead, they argued that it was the low head of the river and surrounding terrain that caused the lower energy output.

The appellate court rejected that reasoning as "specious" and concluded: "It was not merely the failure of the plants, based on their physical characteristics, to generate the desired amount of energy that caused TPI's damage. The legal cause of TPI's injury was Dunlevy's failure to adequately convey the realities of Ogdensburg and Forestport with a level of professional care that would have allowed TPI to make its business decisions with respect to those sites based on reasonably reliable technical information."

The court further stated that the physical traits of the sites served only to provide the conditions precedent to the malpractice committed by Dunlevy. To find "proximate cause" does not require that a jury determine that harm was caused solely by one party, but rather requires only that the identified cause be a substantial factor in bringing about the injury."

Having satisfied itself that the jury did not err in finding proximate cause in the actions of Mr. Dunlevy, the court held, "Once the jury was convinced that the minimal standard of professional care was not met in this case, the proximate cause of TPI's injury could easily be construed to be the carelessness of Dunlevy, and through the doctrine of respondeat superior, his employer, Stetson-Harza."

Economic Loss Rule

Stetson-Harza and Dunlevy argued on appeal that TPI should not have been permitted to recover "lost revenues" since the "economic loss rule" applies in New York. Under that rule, they argued that even if the evidence supported a finding of professional malpractice, the damages awarded were of the type that arise out of breach of contract rather than compensatory damages customarily awarded in tort (negligence) actions.

The basic rule in New York is that where there is no property damage or personal injury, pure economic losses may not be recovered in a negligence action. Only a party in contract with another can seek such purely financial losses, and the recovery is restricted to an action in contract.

In this case, the court concluded that the economic loss rule would not be applied to bar the damages. It held, "While we recognize that some cases have applied the economic loss rule to bar recovery where the only loss claimed is economic in nature [ ], and still others have applied that rule to professional malpractice cases [ ], the better course is to recognize that the rule allows such recovery in the limited class of cases involving liability for the violation of a professional duty. To hold otherwise would in effect bar recovery in many types of malpractice cases."

Expert Testimony Required

On two of the other hydroelectric plants that were at issue, the trial court dismissed TPI's causes of action for breach of contract and malpractice. TPI cross-appealed that aspect of the trial court's judgment. The appellate court affirmed the judgment because TPI had presented no expert evidence concerning professional malpractice at the two sites and consequently the claims lacked proof.

On the question of breach of contract, the court said that TPI's service contracts with Stetson-Harza and Dunlevy did not include a contractual guarantee that any particular site would produce a particular amount of energy. TPI failed to prove any contract breach for those two facilities.

Conflicts of Interest

Dunlevy and HII argued that the trial court erred in allowing evidence of alleged conflict of interest between Dunlevy's duties as an employee of Stetson-Harza and as an owner of HII. They argued that this was highly prejudicial. The appellate court stated that the trial court had broad discretion to admit such evidence.

Negligent Misrepresentation

Another interesting claim that TPI brought against Stetson-Harza and Dunlevy was based on negligent misrepresentation. The trial court dismissed that claim, and the appellate court affirmed the trial court's decision. The basis for the claim was that Stetson-Harza and Dunlevy misrepresented the energy that would be generated from the plants. The trial court found that the alleged misrepresentations related to future events and were promissory in nature rather than factual, and could not support a claim for negligent misrepresentations.

The appellate court agreed, holding, "In the present case, the negligent misrepresentation claim fails because the energy output predictions were mere promises of future output as opposed to present representations of existing fact." But even if they had made representations sufficient to support a claim, the court stated that TPI's claim would fail for lack of reliance on the alleged representations since TPI possessed adequate knowledge that the project's financial success was unlikely and, in any event, "should have known not to rely on the energy output estimates."

Comment

There are several things to learn from this complex case. A design professional may be subject to claims for consequential damages, including lost profits, even in the absence of property damages or personal injury. State law varies on whether the economic loss rule bars recovery for such damages.

Agreements between the parties should specifically address the risks and liabilities to be accepted by each party. Waivers of consequential damages, limitations of liability and similar clauses may be used to better define the commercial relationship between the parties and enable the parties to evaluate whether the risk-to-reward ratio is reasonable.

Caution should be exercised in permitting one individual to serve more than one entity. The question of alleged conflict of interest was apparently raised to the jury in this case and, according to Dunlevy and Stetson-Harza, caused prejudice and harm to their position. It is also interesting to note that despite the lack of performance guarantees, and the fact that the court did not allow the cause of action for negligent misrepresentation to go forward, TPI was able to recover its financial loss on the basis of professional malpractice. From an insurance perspective, this may be a preferable result since a professional liability policy for a design professional would have excluded recovery for losses arising out of non-negligent breaches of warranties and guarantees.

Chapter 25

Surety

25.1 *De Facto* Takeover: Are a Surety's Rights Protected?
25.2 Sureties Walk a Fine Line Between Contractor Default and Claim Investigation

25.1 *De Facto* Takeover: Are a Surety's Rights Protected? (Vol. 7, No. 5)

By: *Michael J. Carrato, Esq.* – Wickwire Gavin, P.C.

A Miller Act surety needs to be aware of certain notice requirements if it decides to take over performance for its principal on a federal construction contract. Informal or "constructive" notice that a surety intends to take over management of a project may not be enough to protect the surety's interest in the contract balance. A recent decision from the United States Court of Federal Claims illustrates the problems which may arise if a surety does not provide the contracting officer with formal notice.

In *American Ins. Co. v. United States*, 62 Fed. Cl. 151 (September 28, 2004), American acted as surety for G&C Enterprises, Inc. on an Air Force construction contract. G&C requested and received progress payments from the Air Force during the course of construction. However, as the project neared the contractual completion date, G&C began to experience difficulties. Concerned that its principal, G&C,

might default, American assumed *de facto* managerial control of the contract and engaged another contractor to complete the work. American did not formally notify the Air Force that it had assumed responsibility for its principal's contractual obligations.

After assuming control of the project, American discovered that G&C had been paid 97% of the contract price by the Air Force while only performing 80% of the work. American filed a suit against the Air Force claiming damages in the amount of $842,000, which represented the difference in the amount paid to G&C by the Air Force and the value of the work it had performed. American argued that it should be equitably subrogated to the United States in the amount of the overpayment.

In granting summary judgment for the Air Force, the Court held that before an obligation arises on the part of the Government to withhold or divert contract funds, the Government must be notified that the surety believes the contractor is in default and cannot complete the contract. Absent such notice, the Government owes no duty to the surety to protect the contract balance. Because the surety may decide that its interests are best served by continuing to have the Government make progress payments to its principal, constructive notice that a contractor has defaulted and that the surety has taken over the performance of the contract is insufficient. Only when the surety becomes a party to the contract with the United States by entering into a takeover agreement with a federal agency does the Government owe the surety any duty. Based on the undisputed facts of the case, American's rights to equitable subrogation never attached. It not only failed to properly notify the Air Force to stop making payments to the contractor, but instead asked the Air Force to continue making payments to G&C.

Further, the Court held that American failed to demonstrate that the Air Force departed from the terms of the contract in making the overpayments to G&C. Under the payment provisions in the contract, the contracting officer was vested with discretion to pay a contractor in excess of the value of the work performed. Pursuant to that language, the contracting officer could use his/her discretion in balancing the Government's interest in proceeding with the contract against possible harm to the surety. The Court found that the contracting officer in this case did not abuse that discretion. The overpayments were made to provide G&C with the cash flow to complete the project. Based on these findings, the Court held that American was not entitled to recover any amount of the overpayments made to G&C.

Comment: According to the *American Insurance* decision, a surety faces a tricky decision when its principal is experiencing financial difficulties. On the one hand, encouraging the Government to continue making progress payments to its principal could help the contractor address its difficulties and successfully complete the contract. However, such payments reduce the funds available to the surety to complete the work in the event its principal defaults. Diligent monitoring by sureties of their principals' performance under their bonded contracts and financial condition is key to avoiding the situation the surety faced in the *American Insurance* case. To the extent a surety is convinced that its principal cannot complete the contract, it should promptly and explicitly notify the contracting officer to stop payments to the contractor. Additionally, once a principal has defaulted, the surety should, with the assistance of counsel, promptly negotiate an appropriate takeover agreement with the Government.

About the Author: Michael J. Carrato was an attorney with the law firm of Wickwire Gavin, P.C., at the time he wrote this article.

25.2 Sureties Walk a Fine Line Between Contractor Default and Claim Investigation (Vol. 5, No. 5)

By: Robert J. MacPherson, Esq.
Postner & Rubin

Surety bonds are contracts and the rights and obligations of the parties will be determined in accordance with basic principles of contract law. The size of the claims and complexity of the project will not impact the result. That is proven by a recent court decision of the New York Court of Appeals decision in *Walter Concrete Constr. Corp. v. Lederle Laboratories, 2003 WL 367460,* and by a decision of the U.S. District Court for the Southern District of New York decision in *United States Fidelity and Guaranty Company, et al. v. Braspetro Oil Services Company, et al.,* 219 F.Supp. 2d. 403 (2002).

Walter Concrete involved a claim by a general contractor ("GC ") against its subcontractor's performance bond surety. The GC had problems with its subcontractor's performance, but never terminated the subcontract. When the subcontractor abandoned the project, the GC did not ask the surety to complete the work. Instead, it completed the subcontractor's work. The GC in turn demanded payment for those costs from the surety. The surety denied the claim, contending that the GC had never notified the surety of its subcontractor's default. The Court of Appeals affirmed a decision granting summary judgment against the surety, finding that the performance bond contained no provision requiring a notice of default as a condition precedent to any legal action against the surety. The only condition imposed on the GC in order to make a valid claim was that it commence its action within the bond's time limits.

In contrast to *Walter Concrete* which involved a commercial building with claims of a little over a half-million dollars, the case in *USF&G v. Braspetro* involved claims by the owner against its contractor's sureties on two related, but distinct projects. One involved a $165 million design-build contract to convert an oil and natural gas platform into a semi-submersible oil and natural gas production platform and was reported to be the largest of its kind ever undertaken. The second project was a $163 million design-build contract to convert an oil tanker into a floating production, storage and offloading vessel.

The bonds at issue in *USF&G* required that the sureties and the contractor be given a pre-default notice and that the sureties' obligations under the bonds would not arise until after the contractor had been declared in default and its contract formally terminated. The owner provided the pre-default notice and eventually formally terminated the contractor's right to proceed with the work, but the sureties disclaimed any liability. They contended that they were prejudiced by the owner's delay in declaring a default.

The court rejected the sureties' claims, finding that after the owner had sent the pre-default notice, the sureties actively discouraged the owner from formally declaring a default. They did so by emphasizing that once a default was declared all work on the projects would stop while the sureties conducted an investigation to determine the propriety of the default. The sureties also told the owner that such an investigation could not begin until a default was declared. According to the Court, this did not stop the sureties from beginning the preparation of their defense to any claims under the bonds. The Court found that when the default was formally declared the sureties made only a token effort to explore the possibility of taking over the contracts and did not conduct a good faith investigation. Rather, they continued their efforts to prepare for litigation while characterizing these activities as an "investigation."

While the Court held that the sureties had no legal obligation to take any specific action prior to a formal declaration of default, once a default was declared they had an obligation to make a good faith investigation into the claims. The court found that the owner was entitled to recover damages of $90 million, plus attorneys' fees on the tanker conversion project. The damage award, plus attorneys' fees, could not exceed the amount of the bond for each project, but the owner was also awarded pre-judgment interest, which is not limited by the amount of the bonds. The sureties have appealed.

Comment: The decision to terminate a contractor and call upon its surety is always difficult. It will invariably cause delay and disruption to the project and involve the expenditure of costs that may never be recovered. Default declarations should never be made lightly and should only be made after exploring all avenues for having the contractor complete the project. This may include a request to the surety that it actively take steps to prevent an impending default. Sureties, however, have no legal obligation to take any affirmative action prior to a declaration of default and will not do so absent compelling reasons. This does not mean that a party faced with a non-performing contractor should refrain from contacting the surety when warning signs appear.

The late submittal of a shop drawing is probably not such a sign, but claims by subcontractors that they have not been paid for work for which the general contractor has been paid most surely is. A bond claimant who has provided complete and accurate information to a surety will be in a better position to insist that the surety act promptly once the default is declared. Claimants must also comply strictly with the notice provisions and any time limitations contained in the bond. Sureties, who under the guise of conducting an investigation into a claimed default, spend their time and resources building a defense, may find their delay in responding exposes them to a greater liability.

About the Author: At the time he wrote this article, Robert MacPherson was an attorney with the nationally recognized law firm of Postner & Rubin, with a practice emphasizing construction law. Mr. MacPherson is now a partner in the New York office of Thelen Reid & Priest, LLP (phone: 212/603-2000, email: rmacpherson@thelenreid.com). This article was originally published in Postner & Rubin's newsletter.

Chapter 26

Time Limitations on Suits

26.1 California Decision Erodes Certainty of 10-Year Statute of Repose against Construction Defect Claims
26.2 Statue of Limitations for Negligence Instead of for Breach of Contract Requires Dismissal of Action against Architect

26.1 California Decision Erodes Certainty of 10-Year Statute of Repose Against Construction Defect Claims (Vol. 7, No. 7)

By: Gregory R. Shaughnessy
Thelen Reid & Priest LLP

For many years, contractors and construction lawyers in California understood that a bright line existed regarding potential liability for construction defects – no liability 10 years after substantial completion of the project. The recent decision by the California Court of Appeal in *Acosta v. Glenfed Development Corp.*, 128 Cal.App. 4th 1278 (2005) took a narrow exception to the 10-year rule, for actions based on "willful misconduct or fraudulent concealment," and expanded the exception to the point that it may swallow the rule. In so doing, the Court took a statute intended to provide certainty and reduce risk and created a great deal of uncertainty and potential added risk.

Background: Code of Civil Procedure §337.15 provides a statute of repose that bars actions to recover damages for construction defects more than 10 years after substantial completion of the work of improvement. Case law has clarified that this 10-year bar does not apply to personal injury claims. *See, Geertz v. Ausonio*, 4 Cal.App.4th 1363 (1992).

In *Acosta*, the court granted summary judgment in favor of a developer/contractor on the grounds that 47 of the 59 named plaintiffs had been brought into the case more than 10 years after recordation of notices of completion on their single-family homes. The plaintiffs argued that an exception to the 10-year rule applied, namely subdivision (f) of §337.15, which provides: "This section shall not apply to actions based on willful misconduct or fraudulent concealment."

In opposition to the motion for summary judgment, counsel for the plaintiffs asserted this exception.

Plaintiffs submitted two declarations from expert witnesses that listed defects that are commonly found in reports by expert witnesses in substantial construction defect litigation, including:

- Horizontal attachments for vertical truss supports in the garages were missing.

- Lack of adequate shear transfer.

- Missing or inadequate holddowns.

- Missing or inadequate straps and hangers on load-bearing members.

- Some driveways were short by as much as 3.5 feet.

- Stucco defects.

- Leaking windows.

- Less expensive kraft paper used to flash around the windows instead of the asphalt polyethelene sheeting specified on plans approved by the city.

The experts opined that most of the defects undermined the structural integrity of the homes and created a substantial risk of injury to persons and/or property. They also opined that the defects:

- Involved conspicuous failures to comply with applicable building code provisions, with the city-approved building plans and with basic construction industry practices.

- Were of a type that inevitably would have been recognized by any competent construction supervisor conducting even minimal day-to-day inspections of the type required in a construction project such as the one at issue and would have caused the construction supervisor to require the responsible subcontractors to remedy the defects immediately before work could proceed on the houses.

- Had the financial impact of producing, in defendants' favor, substantial cost savings.

The experts stated that the defects appeared to be the result of willful misconduct by defendants in that they were "so serious and prevalent that they were either the result of [a] deliberate decision to 'cut corners' for cost savings or the result of a near total, virtually reckless, failure by the developer to adequately supervise subcontractors."

Analysis: The *Acosta* court recognized that other case law had emphasized the purpose of §337.15 was to "provide a 'firm and final' outside limitation period for construction suits involving claims for latent defects." Nonetheless, the court held that the "willful misconduct exception" applied, holding that the term encompassed "not only intentional wrongdoing, but negligence of such a character as to constitute reckless disregard for the rights of others."

The *Acosta* court was able to locate only one other decision applying the exception in §337.15 (f), *Felburg v. Don Wilson Builders*, 142 Cal.App.3d 383, 390 (1983). There, a defendant builder sold plaintiffs a home that had been built over an oil sump. After subsidence caused considerable damage to the home, the plaintiffs filed suit about 12 years after substantial completion of the home. In opposition to the defendant's motion for summary judgment, the plaintiffs submitted an expert declaration that stated that "it would have been impossible to pour the foundation of the home without seeing the evidence, in plain view, that the lot was over an oil sump." The plaintiffs also offered evidence indicating that the builder actually had received a boring report from a soils engineer that showed the existence of the oil sump on or near the plaintiff's lot.

The *Acosta* court found that the facts in *Felburg* were "remarkably similar" to those in the *Acosta* case. It is submitted that the obvious knowledge in *Felburg* that a house was being built on a woefully deficient construction site with total disregard of a soils report showing the existence of the oil sump is not remotely, much less remarkably, similar to garden variety construction defects that were present in the *Acosta* case. Indeed, the principal defects in *Acosta* appeared to be defects that may have undermined the structural integrity of the houses and which created the risk of injury but that had not actually caused any injury.

Importantly, the *Acosta* court held that the developer/contractor could be found to have engaged in willful misconduct even if it did not have actual knowledge of the defects, for example, where the work was performed by subcontractors. The court reasoned that the developer/contractor was liable to buyers for the acts of the subcontractors because developers/contractors "have supervision over the construction, including the work of the subcontractors" and found that this duty was non-delegable. The court also found that imposing supervisory obligations on developers/contractors was consistent with the contractor's license law. Finally, the court found that under the exception in subdivision (f), which states that it applied to "actions based on willful misconduct," "it is only necessary that the action be based on and arise from willful misconduct by someone. It does not matter whether defendants committed such misconduct directly or it was done by subcontractors hired by them."

Comment: The California Supreme Court denied a petition for review and request for depublication. Thus, plaintiffs now have authority that could make overcoming the 10-year statute of repose in most construction defect cases easier. The 10-year statute of repose no longer will be seen as an almost insurmountable barrier.

It is difficult to imagine a case where a creative plaintiff's lawyer will be unable to come up with a declaration that the construction defects were the result of a deliberate decision to cut corners for cost savings and that there must have been a near-complete failure by the developer to exercise even minimal supervision. In such cases, it may be difficult for a defendant to escape from the case by a motion for summary judgment that relies on the 10-year statute of repose. Thus, plaintiffs frequently will be able to get cases before a jury that would otherwise have been disposed of by summary judgment. Whether the conduct by the developer/contractor was willful misconduct normally will be a question of fact for the jury. *Colich & Sons v. Pacific Bell*, 198 Cal.App.3d 1225 (1988).

The *Acosta* decision is not remarkable in its application of the "willful misconduct" exception but rather in the manner in which it applied the exception. It allowed expert declarations that were not remarkable and that did not remotely approach the egregious facts in the *Felburg* decision to create a triable issue of fact, even against a developer/contractor when there was no direct evidence that the developer/contractor had knowledge of the defects. In effect, the *Acosta* decision could create strict liability by the developer/contractor for any willful misconduct by the subcontractors or when the defects were the result of a lack of supervision.

Some steps can be taken by developers/contractors to try and keep the 10-year limitation period intact. This includes conducting special inspections and keeping good records of such inspections. This would tend to show that proper supervision was provided and that attention to quality control was given, perhaps enough to overcome the conclusory and self-serving declarations of the plaintiff's experts on a motion for summary judgment. Notably, there was no discussion in *Acosta* of the fact that the allegedly grossly defective work presumably had passed inspections by local building officials.

Special inspections actually have been fairly common for the last five years on condominium projects in California, as most such projects were built with wrap OCIP insurance policies, which typically require such inspections as part of the OCIP program. However, such OCIP policies typically also have 10-year "tail" coverage following substantial completion. Under *Acosta*, construction defect actions can be brought after the expiration of the 10-year tail, leaving developers, contractors and subcontractors exposed to liability for construction defects with no insurance coverage -

- thus further chilling the market for construction of single-family housing in California.

About the Author: Gregory R. Shaughnessy is an attorney in the law firm of Thelen Reid & Priest in San Francisco. This article originally appeared in a Thelen Reid & Priest newsletter and was published on the firm's website at http://www.ConstructionWebLinks.com. For more information about the issues covered in this report, contact Mr. Shaughnessy in the firm's San Francisco office at 415-369-7235 or at gshaughnessy@thelenreid.com.

26.2 Statue of Limitations for Negligence Instead of for Breach of Contract Requires Dismissal of Action against Architect (Vol. 6, No. 4)

A professional liability claim against an architect was governed by a three-year statute of limitations applicable to non-medical, professional malpractice rather than the six-year statute for actions based on breach of contract. Regardless of whether the alleged failures of the architect were a breach of contract, they arose out of alleged malpractice. Actions to recover damages for malpractice were required by New York law to be commenced within three years regardless of whether the underlying theory is based in contract or negligence.

In this case the allegations were that the architect failed to comply with fireproofing requirements of the Connecticut Building Code applicable to a commercial building being designed and built in Stamford, Connecticut. Almost four years after the building was completed and occupied, the building owner brought a demand for arbitration against the architect. (Although the project was located in Connecticut, the contract apparently specified that New York law would be applied.)

In denying the architect's motion to dismiss the action based on the three year statute of limitations having elapsed, the first court (motion's court) concluded that the plaintiff Owner was entitled to the six-year statute for breach of contract because it was contending in its suit that the architect completely failed to perform its specified contractual responsibility and not that the architect committed malpractice. In reversing that decision, the appellate court stated that "whether [architect's] alleged failure to comply with the applicable code provisions was a breach of contract or tortious [i.e., negligent] in nature is immaterial for statute of limitations purposes, since the resulting non-compliance is the same, as is the remedy sought." The court went on to find that the New York Legislature's intent was that this type of action be subject to the statute of limitations for professional malpractice.

A Legislative Memorandum supporting certain clarifying amendments to the statute of limitations was quoted by the court stating that it was "the legislative intent that where the underlying complaint is one which essentially claims that there was a failure to utilize reasonable care or where acts of omission or negligence are alleged or claimed, the statute of limitation shall ... be three years ... regardless of whether the theory is based in tort or in a breach of contract." *R.M. Kliment & Frances Halsband, Architects v. McKinsey & Company, Inc.*, 770 N.Y.S. 2d 329 (2004 WL 57074 (2004).

Even though the claimant had alleged breach of an express, rather than implied, term of the agreement, the court held that "while compliance with the relevant building code may have been a particular bargained-for result, that result is not inconsistent with an architect's ordinary professional obligations. Making such ordinary obligations express terms of an agreement does not remove the issue from the realm of negligence as argued by McKinsey, nor can it convert a malpractice action into a breach of contract action."

Comment 1: This court concluded that although the architect expressly committed by contract to comply with various code requirements, this did not constitute an express warranty that would turn this into breach of contract claim and thereby avoid requirements applicable to professional malpractice claims—such as the shorter statute of limitations period. This would also mean that the claimant would be required to prove that the architect's alleged lack of code compliance was the result of negligence. Thus, it appears that the issue is whether the architect complied with the generally accepted standard of care. In an action based upon negligence, a claimant would have to prove not only that the architect failed to comply with a code provision, but also that it was negligent in doing so.

Mere failure to comply does not necessarily entitle the claimant to a judgment against the architect. The reasoning of the court is quite important because quite a few contracts seem to be attempting to make the design professional contractually liable for any and all errors regardless of whether there was negligence or not. If the reasoning of this opinion is applied, one might argue just as held here that "Making such ordinary obligations express terms of an agreement does not remove the issue from the realm of negligence ... nor can it convert a malpractice action into a breach of contract action.

Design professionals and their attorneys are sometimes frustrated by some project owners who attempt to convert all professional obligations into strict contractual commitments with the expectation that they can win a claim against the design professional without proving negligence and that they can even prevail to obtain insurance proceeds from the architect's professional liability policy in the absence of negligence. Insurance professionals and risk managers routinely advise design professionals to avoid "contractual liabilities" by which the design professional would have liability for non-negligent performance since the liability expressly excludes coverage for "contractual liability" and express warranties. This decision might be a useful one to share with such owners during contract negotiation. Attempting to hold design professionals responsible for anything other than their negligence is not in the best interest of the project or project owner. As seen in this case, it may be the cause unnecessary confusion and litigation. And it may also cause an insurance carrier to deny coverage for a claim if the owner in fact recovers for breach of contract instead of for negligence.

Comment 2: This case demonstrates the importance of specifying in the contract what law will govern. Will it be the law of the state where the project is built or where the architect is domiciled, or even where the project owner maintains its principal office? The outcome of a case can be dramatically altered by that decision. To clarify when a cause of action accrues for the purpose of measuring the time for filing action, there is much to be said for specifying that date in the contract, for example as the date of substantial completion of construction. The time periods can be further clarified by contractually agreeing to a specified number of years following substantial completion in which a suit or demand for arbitration may be brought. This can avoid completely disputes such as this one over how to interpret state statutes of limitations.

Chapter 27

Warranty

27.1 Contractor Entitled to Rely Upon Government's Implied Warranty of Specifications
27.2 No Warranty of Design by Engineer

27.1 Contractor Entitled to Rely Upon Government's Implied Warranty of Specifications (Vol. 5, No. 1)

Where a contractor had to revise the government's design and expend additional time and expense to construct a door for a helicopter hangar, the government argued unsuccessfully that the contractor was barred by a general disclaimer from claiming entitlement to a change order for its extra costs. The specifications contained a general disclaimer advising prospective contractors to verify the government's design prior to bidding the project. After contract award, the contractor discovered that the specified three-pick-point design for the heavy door would not work. It submitted its own design for a four-pick-point door that was accepted by the government. This resulted in modifications to the door and to trusses supporting the door.

When the government refused to approve a change order, the contractor filed an action with the Armed Services Board of Contract Appeals. The Board held in favor of the contractor, and the government appealed to the Federal Circuit Court of Appeals.

The Court of Appeals affirmed the Board decision in favor of the contractor. The court explained that when the government provides a contractor with design specifications, the contractor is bound by them, and there is an implied warranty that the specifications are free from design defects. [citing *United States v. Spearin*]. As further explained by the court, "This implied warranty attaches only to design specifications detailing the actual method of performance. It does not accompany performance specifications that merely set forth an objective without specifying the method of obtaining the objective. Because the implied warranty protects contractors who fully comply with the design specifications, the contractors are not responsible for the consequences of defects in the specified design."

In this case, the government attempted to avoid the consequences of its implied warranty by shifting the risk to the contractor via a general disclaimer. But the court did not accept this. In fact, the court emphatically stated that general disclaimers that require a contractor to check plans and determine project requirements do not overcome the implied warranty. Only express and specific disclaimers will suffice, said the court, to overcome the implied warranty that accompanies design specifications. In the absence of such disclaimers, a contractor is entitled to any additional costs it reasonably incurs in producing satisfactory results.

Factors that influenced the court in favor of the contractor included several factors. The first factor was that the design of the government was defective but it was not a patent defect that could be readily discovered by a contractor prior to bidding the job. Contractors are not required to investigate to ferret out hidden or subtle errors in the specifications. The second factor was that the government authored the specifications incorporating significant design characteristics that the contractor was required to follow and from which the contractor was not permitted by contract to deviate from without "approval of the Army's architect." The court pointed out that, "If the three-point-pick design had been merely a performance specification ... [contractor] could have chosen any method of building a workable tilt-up canopy door, including a four-pick-point door design." The third factor was that although the disclaimer at issue required the contractor to verify supports, attachments, and loads, it did not clearly alert the contractor that the design may contain substantive flaws, requiring correction and approval before bidding.

Although the general disclaimer did not overcome the implied warranty, it was nevertheless important, said the court. It placed on the contractor the risk to check the accuracy of the physical details provided in the drawings by the government, "but not the design." In summary, the court concluded that the design flaw at issue was hidden and the contractor had no obligation to ferret out the subtle flaw before bidding.

The disclaimer placed responsibility on the contractor for verifying physical details, but it did obligate the contractor to "analyze the Government's design to determine whether it will work for its intended purpose. Since the design details of the specifications created an implied warranty, and since the general disclaimer did not work to overcome that warranty, the contractor was entitled to recover its additional costs incurred in creating a design and constructing a door that worked for its intended purpose. *White v. Edsall Construction Company*, 296 F.3d 1081 (Fed. Cir. 2002).

Comment: This case serves as an excellent reminder that contractors are entitled to rely upon design specifications. Project owners cannot avoid their implied warranty of design by attempting to include vague disclaimers or requirements that the contractor perform pre-bid investigations of its own. The court rightly explained the principles of the *Spearin* doctrine, holding the government responsible for the costs necessary to construct the door to an appropriate design. It is unfortunate to see an increasing number of project owners trying to get around the Spearin doctrine by using onerous contract language against their contractors. In the opinion of this editor, all parties to construction projects are better served by contract language that recognizes entitlement to a change order in circumstances similar to those described by the court in this case.

Efforts by owners to shift the risk to the contractor will inevitably result in an increased number of cost contingencies added to the contractor's bids, and will also logically result in more disputes and contractor claims. Prudent risk management calls for assigning the risk to the party that can best control the risk, and in this case that would appear to be the project owner.

27.2 No Warranty of Design by Engineer (Vol. 5, No. 2)

In response to a plaintiff's suit against an engineer alleging breach of an express warranty, a court held that the complaint must be dismissed because the record revealed no express warranty, and if the service was performed negligently the cause of action must be based in negligence rather than warranty. The engineer had been retained to provide professional services with respect to the design of a sludge treatment facility. In analyzing the allegations contained in the plaintiff's complaint, the court stated that those allegations could not be construed as asserting either a negligence or a breach of contract claim. Citing several New York court decisions, this court stated as a matter of law that "no warranty attaches to the performance of a service." Whatever representations were made by the engineer in its contract did not, in the opinion of the court, rise to the level of guaranteeing a particular result.

Another count of the plaintiff's complaint alleged that the engineer had committed fraud and intentional misrepresentation. The court held that this aspect of the complaint also had to be dismissed because before a plaintiff can pursue such a theory it must first demonstrate that it was prevented by the defendant's fraud from pursuing normal available remedies for negligence or breach of contract. In other words, the plaintiff would have to show that it had been diverted from pursuing those other remedies by its reliance upon the defendant's alleged misrepresentation. Or as explained by the court, "Stated another way, a fraud claim in this regard 'is sustainable only to the extent that it is premised upon one or more affirmative, intentional misreprentations – that is, something more egregious than mere concealment or failure to disclose additional damages, separate and distinct from those generated by the alleged malpractice."

In this case, the court found that the allegations pertaining to intentional misrepresentation and those pertaining to professional malpractice are based on the same factual allegations, namely that the engineer designed a facility "that it knew would not be adequate for its intended purpose and, in so doing, misrepresented the adequacy of the underlying design." The damages sought under both counts of the complaint are also essentially identical. For the foregoing reasons, the appellate court held that the complaint should have been dismissed by the trial court. *Rochester Fund Municipals et al. v. Amsterdam Municipal Leasing Corporation,* 296 A.D.2d. 785, 746 N.Y.S.2d 512 (2002).

Comment: Why is it that project owners, and even contractors, seem increasingly to be throwing the kitchen sink into their complaints against design professionals? As a result of complaints alleging fraud, breach of warranty and breach of performance guarantees, professional liability insurance carriers are either delayed or disabled from quickly resolving the complaints. Instead, they get bogged down for years in arbitration or litigation, and perhaps even worse, the very allegations defeat the ability of anyone to recover under the policy.

Let's think about this rationally. Professional liability policies cover NEGLIGENT acts, errors and omissions. They do not cover intentional misrepresentations and fraud. They do not cover breach of contract that arises out of anything other than negligence. And they do not cover breaches of warranty or guarantees. Consequently, when plaintiffs sue design firms based on any allegations other than negligence, they are throwing a monkey wrench into the works. They are waiving giant red flags to insurance carriers. They are begging for trouble.

Note that despite the general premise stated above, there may be some instances when complaints alleging design firm fraud, bad faith, or contract interference are necessary and appropriate, as when the design firm, without knowledge of the contractor, has signed a contract with the project owner making it responsible for all construction change order costs exceeding the project budget, and the design professional proceeds, as the owner's agent, to deny all change orders and proper claims for equitable adjustment.

What about contract language that creates uninsurable risks? More and more project owners are inserting into their contracts with design firms language obligating the design firm to warrant or guarantee its services, to guarantee cost estimates, to indemnify the owner against all claims and frivolous complaints by anyone, regardless of whether there was any negligence on the part of the design professional. Such language creates uninsurable risk and has the potential to cause much confusion when it comes to analyzing whether there is any insurance coverage available.

If a plaintiff asserts that the design firm has liability based upon a contractual obligation that is not related to negligence, the insurance carrier will decline coverage and will likely even decline to defend the case where there is no potential recovery available under the terms of the policy. So, why do owners do this? I believe it is largely because they are being misguided and misadvised by well meaning, but sometimes ill-informed attorneys.

Insurance companies like DPIC, CNA/Schinnerer, Zurich, and ARCH have been consistently, and for many years, providing contract reviews, giving lectures, teaching workshops, and writing articles and papers, explaining that the insurance covers negligence only and that the contracts need to avoid creating uninsurable contractual liability and warranties. With all this information out there, does there not become a standard of care applicable to those representing project owners on construction projects to know how professional liability insurance works and to advise their clients accordingly so as to facilitate the recovery of insurance proceeds under the policy?

Despite all of this, we continue to read about, and deal with, plaintiffs that waste time and resources by filing complex complaints filled with uninsurable claims, when they could make their lives so much easier by keeping their contracts and their litigation focused on the one thing that matters most, and that is whether the design professional committed a NEGLIGENT act, error or omission. We even have associations of owners that are creating their own standard form contracts that would create uninsurable risks for design firms along the lines described above.

Having reached a point of genuine frustration with the direction we seem to be heading with contracts and litigation, I am perhaps taking a stronger stand than might be politically correct. But I think it is time to take a stand.

Index

Acceleration
Proving Constructive Acceleration Claims Can be Difficult for Contractors, 4.6

Accord & Satisfaction
Accord and Satisfaction Barred Contractor Claim for Additional Compensation, 1.1
Accord and Satisfaction Language Barred Contract Claim, 1.2
Do You Really Want to Cash that Check?, 1.4
Waiver and Release in Change Order Bars Further Recovery, 1.3

Americans with Disabilities Act
Design Professionals Not Subject to Liability under Title III of the ADA and Washington's Law against Discrimination, 2.1

Arbitration
Arbitration Consolidation was Inappropriate, 8.5
Architect's Decision Final Where Contractor Failed to Satisfy Arbitration Filing Requirements, 8.6
Subcontractor Forfeits Right to Arbitration by Filing Demand Untimely, 8.2
Using Negotiation, Mediation and Arbitration to Resolve Construction Disputes, 8.3
Why Some Mediations Fail, 8.4

Architect's Decisions on Claims
Architect's Decision Final Where Contractor Failed to Satisfy Arbitration Filing Requirements, 8.6

Changes
Equitable Adjustment Allowed for Deductive Change Despite Contractor's Unbalanced Bid, 4.5
Managing Contract Changes, 3.1

Claims (See Contractor Claims)
Architect's Decision Final Where Contractor Failed to Satisfy Arbitration Filing Requirements, 8.6

Contract Language Issues & Concerns
Boilerplate Can Burn!, 5.1

Contractor Claims
Absent Contractual Allocation of Delay Risk, Subcontractor Cannot Recover Damages from General Contractor who was not Responsible for their Occurrence, 4.4
Contractor Suit Dismissed for Failure to Follow Claim Procedures of Contract, 4.1
Contractors May Now Bring Direct Action for Economic Losses against Design Professionals in Pennsylvania, 11.1
Contractor Complaint against Engineer Dismissed for Failure to File Expert Identification Affidavit, 14.1
Differing Site Conditions, Defective Specifications: One Coin, Two Sides, 4.8
Equitable Adjustment Allowed for Deductive Change Despite Contractor's Unbalanced Bid, 4.5
Failure to Request Change Order Bars Contractor Recovery for Excess Units, 4.2
Multi-Million Dollar Claim Invalidated by Court Due to Contractor's Failure to Give Timely Notice, 4.3
Proving Constructive Acceleration Claims Can be Difficult for Contractors, 4.6
When to Stop Work for Non-payment, 4.7

Competitive Bidding
Public Agency Exempted Project from Competitive Bidding, 7.3

Copyright
Copyright Infringement of Design Documents, 9.2

Damages
Contractor May be Sued for Lost Profits arising out of Breach of Contract, 6.1
Liquidated Damages Clause and Waiver of Consequential Damages Clause Effectively Cap Damages Available Against Design-Builder, 7.2
Punitive Damage Award Against Insurance Company Reversed by Supreme Court as Excessive, 19.6

Delay and Impact Claims
Absent Contractual Allocation of Delay Risk, Subcontractor Cannot Recover Damages from General Contractor who was not Responsible for their Occurrence, 4.4
Managing Contract Changes, 3.1

Design-Build
Design-Builder Not Entitled to Equitable Adjustment to Meet Owner's Detailed Design Specifications, 7.1
Design-Build Engineer Held Liable for Negligence, 24.3
Liquidated Damages Clause and Waiver of Consequential Damages Clause Effectively Cap Damages Available Against Design-Builder, 7.2
Public Agency Exempted Project from Competitive Bidding, 7.3
Superfund Decision May Benefit Design Professionals on Environmental Remediation Projects, 12.1

Differing Site Conditions
Differing Site Conditions, Defective Specifications: One Coin, Two Sides, 4.8

Dispute Resolution
Arbitration Consolidation was Inappropriate, 8.5
Architect's Decision Final Where Contractor Failed to Satisfy Arbitration Filing Requirements, 8.6

Contractual Jury Waivers Held Invalid by California Supreme Court, 8.1
Subcontractor Forfeits Right To Arbitration By Filing Demand Untimely, 8.2
Using Negotiation, Mediation and Arbitration to Resolve Construction Disputes, 8.3
Why Some Mediations Fail, 8.4

Documentation
Copyright Infringement of Design Documents, 9.2
Don't Touch That "Forward" Button! Attorney-Client Privilege in an E-Mail Age. 9.1

Drug Testing
Rapid Result Drug Testing, 10.1

E-Mail
Don't Touch That "Forward" Button! Attorney-Client Privilege in an E-Mail Age. 9.1

Economic Loss Doctrine
Contractors May Now Bring Direct Action for Economic Losses against Design Professionals in Pennsylvania, 11.1

Environmental Liability
Pollution Exclusion in D&O Policy Applied to Exclude Coverage for Alleged Business Torts, 19.3
Superfund Decision May Benefit Design Professionals on Environmental Remediation Projects, 12.1

Ethics
Testing Your Ethical Barometer, 13.1

Expert Witnesses
Contractor Complaint against Engineer Dismissed for Failure to File Expert Identification Affidavit, 14.1
Personal Injury Case against Engineer Dismissed for Lack of Expert Testimony, 14.2
Summary Judgment Against Plaintiff Who Failed to Provide Factual Support That She Could Meet Burden of Proof of Negligent Design, 24.2

Faulty Workmanship
Faulty Workmanship Coverage Under CGL Policy, 19.2

Federal Contracts
Hurricane Katrina's Impact on Existing U.S. Government Contracts, 15.1

Fiduciary Duty
Project Manager Required by Fiduciary Duty to Owner to Agree to Settlement with a Supplier Contrary to its Own Interest, 16.1

Indemnification
Highway Contractor Protected by State Immunity Statute, 17.2
Indemnity Clause Requires Subcontractor to Indemnify Prime for Injuries Arising out of the Prime's Own Negligence, 17.1
Indemnification Clause Unenforceable if Negligent Parties Are Indemnified, 17.3

Jury Waivers
Contractual Jury Waivers Held Invalid by California Supreme Court, 8.1

License Requirements
Contractor Forfeited Right to Payment by Performing Work without a License, 18.1

Liquidated Damages
Liquidated Damages Clause and Waiver of Consequential Damages Clause Effectively Cap Damages Available against Design-Builder, 21.2

Insurance
Broad Additional Insured Endorsement Entitles Contractor to Recover Damages under its Subcontractor's Primary and Umbrella Policies, 19.5
Faulty Workmanship Coverage Under CGL Policy, 19.2
Insurance Coverage—Waivers of Subrogation, 19.8
Insurance Carrier not Required to Treat CM as Additional Insured Under Contractor's Policy, 19.4
Insurance Company that Incorrectly Denied Pollution Coverage Did Not Act in Bad Faith in Failing to Defend and Indemnify its Insured, 19.7
Pollution Exclusion in D&O Policy Applied to Exclude Coverage for Alleged Business Torts, 19.3
Punitive Damage Award Against Insurance Company Reversed by Supreme Court as Excessive, 19.6
Waiver of Subrogation Enforced, Denying Insurance Company Recovery against Contractor it Alleged was Grossly Negligent, 19.1

Insurance Coverage for Environmental Losses & Mold
Absolute Pollution Exclusion in Contractors Policy Does Not Bar Coverage for Injuries from Toxic Fumes, 20.6
Broad Pollution Exclusion Is Ambiguous: Lead Covered by Policy, 20.2
Homeowners Policy Unambiguously Excluded Coverage for Mold, 20.5
Mold Loss Excluded under Homeowner's Policy – Summary Judgment for Carrier, 20.4
Silica Claim Barred by Total Pollution Exclusion in CGL Policy, 20.1
Whether Mold Cleanup Costs Are Covered Depends on Causation, 20.3

Pollution Exclusion in D&O Policy Applied to Exclude Coverage for Alleged Business Torts, 19.3

Limitation of Liability
Limitation of Liability Clause Protecting Owner is Not Voided by Owner's Breach of Contract or Alleged Bad Faith, 21.1
Liquidated Damages Clause and Waiver of Consequential Damages Clause Effectively Cap Damages Available against Design-Builder, 21.2

Mediation
Using Negotiation, Mediation and Arbitration to Resolve Construction Disputes, 8.3
Why Some Mediations Fail, 8.4

Mold
Court Rejects Employees' Claim against Employer for Fraudulent Concealment of Mold, 22.3
Incident Reports are Held to be privileged, 22.4
Homeowners Policy Unambiguously Excluded Coverage for Mold, 22.5
Preventing Mold-Related Nondisclosure Claims, 22.2
Standards Needed for Mold Exposure, Testing and Remediation, 22.1
Whether Mold Cleanup Costs Are Covered Depends on Causation, 20.3

Negligence
Architect Required to Review Adequacy of Engineer's Structural Report Before Proceeding with its Design Services, 24.1
Design-Build Engineer Held Liable for Negligence, 24.3
Indemnity Clause Requires Subcontractor to Indemnify Prime for Injuries Arising out of the Prime's own Negligence, 17.1
Indemnification Clause Unenforceable if Negligent Parties Are Indemnified, 17.3

Summary Judgment against Plaintiff who Failed to Provide Factual Support That She Could Meet Burden of Proof of Negligent Design, 24.2

Notice Requirements
Contractor Suit Dismissed for Failure to Follow Claim Procedures of Contract, 4.1
Failure to Request Change Order Bars Contractor Recovery for Excess Units, 4.2
Multi-Million Dollar Claim Invalidated by Court Due to Contractor's Failure to Give Timely Notice, 4.3
Subcontractor Forfeits Right to Arbitration by Filing Demand Untimely, 8.2

Payment Disputes
Contractor Forfeited Right to Payment by Performing Work without a License, 18.1
When to Stop Work for Non-payment, 4.7

Project Manager Duty
Project Manager Required by Fiduciary Duty to Owner to Agree to Settlement with a Supplier Contrary to Its Own Interest, 16.1

Settlement
Project Manager Required by Fiduciary Duty to Owner to agree to Settlement with a Supplier Contrary to Its Own Interest, 16.1

Site Safety
Engineer Had No Duty to Warn General Contractor's Employee of Danger, 23.2
No Liability under New York Labor Law for Project Owner and Lender Where Worker's Injuries Attributed Solely to His Own Fault, 23.3
Store Owner Not Liable for Injuries Sustained by HVAC Contractor's Employee, 23.1

Subcontractor Obligations
Indemnity Clause Requires Subcontractor to Indemnify Prime for Injuries Arising out of the Prime's Own Negligence, 17.1

Standard of Care (see also Negligence)
Architect Required to Review Adequacy of Engineer's Structural Report Before Proceeding with its Design Services, 24.1
Design-Build Engineer Held Liable for Negligence, 24.3
Summary Judgment against Plaintiff Who Failed to Provide Factual Support That She Could Meet Burden of Proof of Negligent Design, 24.2

Specifications
Differing Site Conditions, Defective Specifications: One Coin, Two Sides, 4.8

State Immunity
Highway Contractor Protected by State Immunity Statute, 17.2

Statute of Repose
California Decision Erodes Certainty of 10-Year Statute of Repose against Construction Defect Claims, 26.1

Stop Work
When to Stop Work for Non-Payment, 4.7

Surety
De Facto Takeover: Are a Surety's Rights Protected? , 25.1
Sureties Walk a Fine Line Between Contractor Default and Claim Investigation, 25.2

Time Limitations on Suits
California Decision Erodes Certainty of 10-Year Statute of Repose against Construction Defect Claims, 26.1

Framing Professional Negligence Claim as Breach of Contract Does Not Get Around Shorter Statute of Limitations Period for Negligence Claims, 26.2

Statue of Limitations for Negligence Instead of for Breach of Contract Requires Dismissal of Action against Architect, 26.3

Waiver of Consequential Damages

Liquidated Damages Clause and Waiver of Consequential Damages Clause Effectively Cap Damages Available against Design-Builder, 21.2

Waiver of Subrogation

Waiver of Subrogation Enforced, Denying Insurance Company Recovery against Contractor It Alleged Was Grossly Negligent, 19.1

Warranty

Contractor Entitled to Rely Upon Government's Implied Warranty of Specifications, 27.1

No Warranty of Design by Engineer, 27.2